Death by Dogma:
The biological reason why the Left is leading us to extinction, *and the solution*

Jeremy Griffith

www.HumanCondition.com

Death by Dogma: The biological reason why the Left is leading us to extinction, and the solution by Jeremy Griffith

Published in 2021, by WTM Publishing and Communications Pty Ltd (ACN 103 136 778) (www.wtmpublishing.com). Revised 2022.

 All enquiries to:
WORLD TRANSFORMATION MOVEMENT®
Email: info@worldtransformation.com
Website: www.humancondition.com or www.worldtransformation.com

The World Transformation Movement (WTM) is a global not-for-profit movement represented by WTM charities and centres around the world.

ISBN 978-1-74129-066-0
CIP – Biology, Philosophy, Psychology, Health

Contents

Background

Jeremy Griffith is an Australian biologist who has dedicated his life to bringing fully accountable, biological understanding to the dilemma of the human condition—the underlying issue in all human life of our species' extraordinary capacity for what has been called 'good' and 'evil'.

Jeremy has published seven books on the human condition, including the Australasian bestseller *A Species In Denial* (2003), his definitive treatise, *FREEDOM: The End Of The Human Condition* (2016), and in 2020 *THE Interview*, the transcript of acclaimed British actor and broadcaster Craig Conway's astonishing, world-changing and world-saving interview with Jeremy about his book *FREEDOM*. Since Critical Theory wasn't taking hold in society when *FREEDOM* was published in 2016, in 2021 Jeremy wrote this companion booklet for both *FREEDOM* and *THE Interview* to explain the extreme threat it presents to humanity.

His work has attracted the support of such eminent scientists as the former President of the Canadian Psychiatric Association Prof. Harry Prosen, Australia's Templeton Prize-winning biologist Prof. Charles Birch, the Former President of the Primate Society of Great Britain Dr David Chivers, Nobel Prize winning physicist Stephen Hawking, as well as other distinguished thinkers such as Sir Laurens van der Post.

Jeremy is the founder and patron of the World Transformation Movement (WTM)—see www.humancondition.com.

Jeremy Griffith in 2020

Death by Dogma:
The biological reason why the Left is leading us to extinction, *and the solution*

1. Summary

[1]The Left's dogmatic insistence that everyone behave in a cooperative and loving way makes its advocates feel good but it oppresses and stifles the freedom of expression and individualism needed to find knowledge, ultimately self-knowledge, the redeeming understanding of the human condition that *actually* brings about a cooperative and loving world. Dogma is not the cure, it's the poison because it blocks the search for the rehabilitating understanding of ourselves that's needed to actually save the world. George Orwell's famous prediction that **'If you want a picture of the future, imagine a boot stamping on a human face** [the human mind] **for ever'** was about to come true — but, mercifully, science has finally made it possible to explain the human condition and so save us from this makes-you-feel-good-but-is-actually-horrifically-selfish-and-deluded left-wing threat of the *Death by Dogma* extinction of our species! (Note, a powerful, slightly longer, illustrated summary of this book can also be found in Addendum 1 at the end of the book.)

2. Introduction

[2]To present the biological explanation for why the left-wing's postmodern, deconstructed, politically correct, Marxist, woke, critical theory culture is rapidly leading to our species' extinction, and not to utopia as its supporters delude themselves it will, I first need to briefly summarise what I explain in *THE Interview (TI)* at

www.humancondition.com. <u>*TI* has proved to be such a good intro-</u><u>duction to the explanation of the human condition that it is the ideal</u><u>starting point for any analysis of the human condition—and why</u><u>it needs to be listened to or read before reading this book *Death by*</u><u>*Dogma*</u>. Indeed, Professor Harry Prosen, the former President of the Canadian Psychiatric Association, described *TI* as **'the most important interview of all time because it turns all the conventional knowledge about human existence on its head with its recognition of the original cooperative and loving innocence of our species'.**

[3]Yes, being able to at last explain the good reason why we corrupted our species' original all-loving and all-sensitive, innocent instinctive self or soul (which is essentially what *TI* presents) gives us the ability to at last admit that our species was once innocent, cooperative and loving, and what this does is bring about such a fundamental change in all our thinking that virtually all our knowledge has to be re-presented in a whole new way. Basically, all the dishonest denial the human race has been employing to cope with the until now unbearable truth of our corrupted, 'fallen', good-and-evil-afflicted condition goes, and since that denial has been central to all our thinking, all our thinking suddenly changes. As we are about to see, a whole new truthful reconciled, redeemed and rehabilitated world suddenly opens up, and everything about our existence becomes clear.

3. The key concepts explained in *THE Interview*

[4]I began *TI* by pointing out that virtually everyone is living with the belief—well, it's what we were all taught at school and is reinforced in every documentary—that humans' competitive, selfish and aggressive behaviour is due to us having savage, must-reproduce-our-genes instincts like other animals have. We make comments like: 'We are programmed by our genes to try to dominate

others and be a winner in the battle of life'; and 'Our preoccupation with sexual conquest is due to our primal instinct to sow our seeds'; and 'Men behave abominably because their bodies are flooded with must-reproduce-their-genes-promoting testosterone'; and 'We want a big house because we are innately territorial'; and 'Fighting and war is just our deeply-rooted combative animal nature expressing itself'. I did mention that while left-wing thinkers do claim we have some selfless, cooperative instincts, they also say we have this selfish, competitive 'animal' side, which, as we will see later, Karl Marx limited to such basic needs as sex, food, shelter and clothing. (paragraphs 15-16 of *TI*—please note that early transcripts of *TI* don't have the paragraphs numbered)

[5]I then explained that this was obviously a false excuse for our selfish, competitive and aggressive human condition because words used to describe our behaviour such as egocentric, arrogant, inspired, depressed, deluded, pessimistic, optimistic, artificial, hateful, cynical, mean, sadistic, immoral, brilliant, guilt-ridden, evil, psychotic, neurotic and alienated, all recognise the involvement of OUR species' fully conscious thinking mind. They demonstrate that there is a *psychological* dimension to our behaviour; that we don't suffer from a genetic-opportunism-driven 'animal condition', but a conscious-mind-based, *psychologically* troubled HUMAN CONDITION. Further, I pointed out that humans have selfless, cooperative and loving *moral* instincts, the voice or expression of which we call our conscience—which is the complete *opposite* of selfish, competitive and aggressive instincts. I quoted Charles Darwin saying that **'The moral sense...affords the best and highest distinction between man and the lower animals'** (*The Descent of Man*, 1871, ch.4). And I pointed out that to have acquired our selfless, cooperative and loving moral instinctive self or soul our distant ape ancestors must have *lived* selflessly, cooperatively and lovingly. Our ape ancestors can't have been brutal, club-wielding, selfish, competitive and aggressive savages as we have been taught, rather they must have lived in a Garden

of Eden-like state of selfless, cooperative and loving innocent gentleness. I explained that this selfless, cooperative and loving state was developed through the nurturing of their infants, a process that the bonobo species of ape is currently developing to create their extraordinarily cooperative, loving and gentle behaviour. And I then pointed out that anthropological findings now evidence that our ape ancestors did once live in a cooperative, harmonious state, with, for example, anthropologist C. Owen Lovejoy reporting that **'our species-defining cooperative mutualism can now be seen to extend well beyond the deepest Pliocene** [which is well beyond 5.3 million years ago]' ('Reexamining Human Origins in Light of *Ardipithecus ramidus*', *Science*, 2009, Vol.326, No.5949). (pars 19-20)

Our ape ancestors were innocent, loving, nurturers,

Paleoartist Jay H. Matternes's unusually honest reconstruction of our ancestor, the 4.4 mya *Ardipithecus ramidus*, which appeared in the Dec. 2009 edition of *Science* — see Freedom Essay 22 for more on the fossil evidence of our nurtured past.

NOT savage, barbaric brutes as they have for so long been portrayed.

It is us humans now who are psychotic angry, egocentric and alienated, seemingly 'evil' monsters!

Detail from Jean-Michel Basquiat's 1982 'Untitled' painting which was sold in May 2017 for US$110.5 million, which, at the time was the sixth most expensive artwork ever sold at auction, no doubt because of its extraordinarily honest portrayal of the true nature of our present horrifically psychologically upset human condition—see Freedom Essay 30.

[6] With regard to being taught at school and told in documentaries that we have selfish, savage, competitive and aggressive, must-reproduce-our-genes instincts, I mentioned that some left-wing thinkers claim that along with selfish, competitive instincts, we do have selfless cooperative instincts. I will explain the dishonest theory that the Left used to supposedly account for how we acquired these selfless instincts more fully later, but the following provides a very brief description.

[7] To counter the selfishness-justifying, individualistic, right-wing assertions that we humans are naturally selfish because we have selfish, savage, 'survival of the fittest', must-reproduce-our-genes instincts (which we actually don't have because we have nurtured unconditionally selfless moral instincts), the Left developed equally dishonest selflessness-emphasising theories that maintain that while we do have some selfish, savage, 'survival of the fittest', must-reproduce-our-genes instincts, we also have some selfless instincts derived from situations where supposedly cooperation proved a more successful survival strategy than competition. This so-called 'group selection' theory for the existence of cooperative instincts within us

is false biology because of 'the tendency of each group to quickly lose its altruism through natural selection favoring cheaters [selfish, opportunistic individuals]' ('Can Darwinism improve Binghamton?', *The New York Times*, 9 Sep. 2011), as the biologist Jerry Coyne pointed out. The reality under natural selection is, 'By all means you can help me reproduce my genes but I'm not about to help you reproduce yours'; it was only the extended nurturing of our infants that could overcome genetic selfishness and develop unconditionally selfless behaviour (see Freedom Essay 21). So the right-wing used the false argument that we have savage, must-reproduce-our-genes instincts to justify selfish individualism, and the Left used the equally dishonest argument that we also have group-selected selfless instincts to justify communalism. As is also going to be explained later, left-wing thinkers eventually moved on from the seemingly irreconcilable right-wing biological view that 'Selfishness is natural' versus the left-wing biological view that 'No, selflessness is natural', to a Marxist-based 'Bypass biology, just dogmatically impose selfless, cooperative, social, communal values.' And it is this latter, outrageously dishonest and extremely dangerous so-called Critical Theory that has, since about 2020, been fast replacing the 'Selfishness is only natural because it's in our genes' thinking that has prevailed to date. Yes, what is being taught in schools and presented in documentaries is certainly changing rapidly, and what is going to be explained in this book is just how incredibly false and dangerous this Critical Theory dogma is. In fact, it's so dangerous it's threatening to take our species to a death by dogma extinction!

[8] What will now be described is the *true* explanation for how our original nurtured, completely cooperative, selfless and loving instinctive self (not completely selfish as the right-wing have claimed, or partially selfless and partially selfish as the left-wing has claimed, or neither selfish or selfless as Critical Theorists claim) became corrupted by the emergence in us of competitive and aggressive behaviour when we became a fully conscious species some 2 million years ago.

[9] In *TI* I used my Adam Stork analogy to present this fully accountable and thus true, instinct vs intellect explanation of our corrupted human condition, where our instinctive orientations unjustly condemned our self-adjusting conscious mind's need to understand cause and effect; in other words, to search for knowledge. I explained that this conflict between our instincts and intellect unavoidably caused us to become psychologically upset, seemingly evil angry, egocentric and alienated monsters.

[10] Basically, we humans had no choice but to heroically persevere with the upsetting search for knowledge until we found the particular knowledge we needed, which was the redeeming and healing biological explanation for *why* we corrupted our original all-loving and all-sensitive instinctive self or soul. Until we found that relieving understanding all we could do was continue searching for it, and suffer becoming more and more psychologically upset. In the words from the song *The Impossible Dream* from the musical the *Man of La Mancha*, we had to be prepared to **'march into hell for a heavenly cause'** (lyrics by Joe Darion, 1965); we had to lose ourselves to find ourselves; we had to suffer becoming angry, egocentric and alienated until we found sufficient knowledge to explain ourselves and by so doing rehabilitate and end all our upset. (pars 29-43)

[11] As I emphasised in *TI*, what this real instinct vs intellect explanation of the human condition fundamentally does is lift the 'burden of guilt' from the human race for destroying the magic world of our soul. It establishes that we humans are good and not bad after all. While we are all inevitably variously angry, egocentric and alienated from our different encounters with humanity's heroic battle to find knowledge, ultimately self-knowledge, understanding of our corrupted condition, we can now fully understand and know that every human is equally fundamentally good, special and wonderful. In fact, we are all not just good, we are the heroes of the whole story of life on Earth because the conscious mind is surely nature's greatest invention, and we had to champion its goodness over the ignorance of instincts. Far from being evil, sinful, guilty monsters, we humans

are sublimely wonderful, truly divine beings! Best of all, this ability
to understand and know there was a good reason why the human race
became psychologically upset, is <u>the key, relieving understanding
we have been in search of</u> ever since we became conscious some
2 million years ago and our soul-corrupted condition emerged. As
the psychoanalyst Carl Jung said, **'wholeness for humans depends on the
ability to own our own shadow'**, and since we can now **'own'** the **'shadow'**
of our species' 2-million-year-corrupted condition, the human race
is finally in a position to become **'whole'**. (pars 106-107)

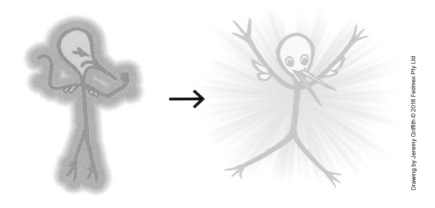

[12] I also described how most of our mythologies and all of our
religions recognised that we humans did once live in a selfless,
cooperative and loving innocent state, and also that it was when
we became conscious that everything seemed to go wrong and we
became the brutally competitive and aggressive, divisively behaved
people we are today. I referred to the examples of Moses's Garden
of Eden story, and to the writings of ancient Greek thinkers Hesiod
and Plato. (pars 52-61)

[13] Crucially though, until science found understanding of the
difference between naturally selected orientating genetic instincts
and nerve-based, memory-capable, cause-and-effect-understanding
consciousness it wasn't possible to explain the good reason for why

we became psychologically upset and soul-corrupted when we became conscious. So while mythologies were able to recognise that becoming conscious somehow caused the corruption of our instinctive self or soul, without science's understandings of the difference in the way gene-based instincts and the nerve-based conscious intellect work, it wasn't possible to explain *why* becoming conscious corrupted our original, gentle and loving, non-conscious instinctive self or soul. (pars 51, 53, 61, 63)

[14] I then made the point that the more we corrupted our all-loving and all-sensitive, but non-conscious, innocent instinctive self or soul, the more difficult it became to admit the truth that our species did once live in such a wonderful state. The more upset we became, the more shame we felt, and thus the more we had to hide from the truth of our species' original state of loving innocence. Hiding in Plato's famous metaphorical dark cave of denial where no exposing light of truth could reach us, saved us from unbearable condemnation. (Plato's cave allegory is described in Video/Freedom Essay 11.) Only a very rare few individuals who were fortunate enough to escape encountering upset during their infancy and childhood could dare face the truth of our species' corrupted human condition and thus think truthfully about it. For everyone else, hiding from the issue of the human condition was paramount.

Computer graphic by James Press © 2018 Fedmex Pty Ltd

[15] Yes, when the great poet Gerard Manley Hopkins wrote about how unbearably depressing the subject of the human condition could be in his aptly titled poem *No Worst, There Is None* (1885), his words, **'O the mind, mind has mountains; cliffs of fall, frightful, sheer, no-man-fathomed'**, did not exaggerate the depth of depression virtually all humans faced if they allowed their minds to think about our soul-corrupted condition while it was still to be **'fathomed'**/understood. The comedian Rod Quantock certainly wasn't joking when he said, **'Thinking can get you into terrible downwards spirals of doubt'** ('Sayings of the Week', *The Sydney Morning Herald*, 5 Jul. 1986). The Nobel Laureate Albert Camus wasn't overstating the problem either when he wrote that **'Beginning to think is beginning to be undermined'** (*The Myth of Sisyphus*, 1942); nor was another Nobel Prize winner, Bertrand Russell, when he said, **'Many people would sooner die than think'** (Antony Flew, *Thinking About Thinking*, 1975, p.5 of 127). And nor was yet another Nobel Laureate for Literature, the poet T.S. Eliot, when he wrote that **'human kind cannot bear very much reality'** (*Burnt Norton*, 1936). Bono, of the rock band U2, sang about how it was better to not just not think at all, but even be deaf, dumb and blind to the issue of the human condition: **'It's been a long hot summer, let's get under cover, don't try too hard to think, don't think at all. I'm not the only one staring at the sun, afraid of what you'd find if you take a look inside. Not just deaf and dumb, I'm staring at the sun, not the only one who's happy to go blind'** (*Staring At The Sun*, 1992). The great Spanish artist Francisco Goya also acknowledged the true pain of the human condition when he created the following picture with the words written on it **'The sleep of reason** [letting down our mental guard] **brings forth monsters'**. (F. Essay 30 describes the excruciating agony adolescents have had to go through learning to resign themselves to living in denial of the human condition; what a relief it will be for adolescents now that understanding of the human condition has been found to no longer have to go through this horrific process of Resignation.)

Goya's *The sleep of reason brings forth monsters*, 1796-97

[16] And so when the upset, soul-corrupted condition developed in humans to the point where trying to face the truth of the extent of it had become unbearable for virtually everyone, a desperate need arose to contrive *some* excuse for why we had turned utopia into dystopia. And it was at that point that we came up with the patently false excuse that I opened *TI* with, which is that our ancestors were not gentle, cooperative and loving innocents, but aggressive savages engaged in relentless competition to reproduce their genes like other

animals. As I emphasised in *TI*, false as it is, it was an absolutely brilliant excuse, because instead of our instincts being all-loving and thus unbearably condemning of our present non-loving state, they were made out to be the vicious and brutal must-reproduce-your-genes instincts seen in other animals; *and*, instead of our conscious mind being the instinct-defying cause of our corruption, which it actually was, it was made out to be the blameless mediating 'hero' that had to step in and try to control those supposed vicious instincts within us! We relievingly found a way to make ourselves out to be heroes, which we actually are, but the argument being employed to achieve it was one enormous 'reverse of the truth' lie! (pars 63-64)

[17] The problem for science, our designated vehicle for enquiry, was that to find the true instinct vs intellect explanation of the human condition required confronting and thinking truthfully about our corrupted condition, which, as explained, has been impossible for virtually all humans, which therefore includes virtually all scientists. So while human-condition-avoiding, so-called 'mechanistic' scientists could do all the hard work of finding the crucial details we needed to explain ourselves, of how genes can orientate but only nerves can understand, not being able to confront and think truthfully about the human condition meant they weren't in a position to assemble that truthful instinct vs intellect explanation. The assemblage of that redeeming truth required exceptional innocence because only it was sufficiently unafraid of the issue of the human condition to confront the issue and thus think truthfully about it, which is how the great philosopher Sir Laurens van der Post and I were able to explain the human condition. (Sir Laurens van der Post's immense contribution to understanding the human condition is explained in F. Essay 51.) Truthful thinking was needed at the end, but all the preparatory hard work was carried out by mechanistic scientists. (pars 70-71)

[18] The issue now that the true, redeeming instinct vs intellect explanation of our corrupted condition has finally been assembled,

and thus the need to use the false 'savage instincts' excuse has ended, is that most scientists are still so habituated to living in fearful denial of the human condition that they find it difficult even reading about the human condition, let alone absorbing its explanation enough to realise the human condition has finally been explained and made safe to confront. This is the 'deaf effect' stage that has to patiently be negotiated before the world realises we humans are finally liberated from our agonising condition. The fabulous dawn of our species' emancipation from our corrupted condition has come, but it is going to take a little while for the warmth of those rays of redeeming truth to thaw us all out! (pars 71-74)

[19] The remainder of *TI* elaborates on this point, on how everyone can now leave Plato's horrible dark cave of psychotic and neurotic lies and delusion, and, as was yearned for in the 1960s musical *Hair*, we can **'Let the sun shine in'** and achieve **'the mind's true liberation'**!

4. Humanity's harrowing journey of ever-increasing denial and artificial relief and delusion to cope with the ever-increasing corruption of our original instinctive self or soul

[20] Having summarised what was presented in *THE Interview*, I am now in a position to explain in some detail the biological explanation for why the left-wing's postmodern, deconstructed, politically correct, Marxist, woke, critical theory culture is rapidly leading to our species' extinction.

[21] As I wrote at the start of this book, what is going to be described is a harrowing 2-million year journey (for, as mentioned, that is how long we have been fully conscious) of ever-increasing artificial relief, delusion and denial to cope with ever-increasing levels of psychological upset, which has taken us to where we are today on the brink of total madness and the imminent threat of extinction.

[22](What is presented here is largely drawn from chapters 8:16 H-Q of my book *FREEDOM*, although the pseudo emancipation ideology of Critical Theory wasn't addressed in *FREEDOM* because it wasn't taking hold over society when *FREEDOM* was published in 2016.)

[23] As the Adam Stork analogy explains, the first form of artificial relief our conscious self-managing mind engaged in when it was unjustly condemned by our instinctive self for defying it was to attack the criticism (anger), deny and block out the criticism (alienation/psychosis), and find any positive reinforcement we could to relieve us of the condemning criticism (egocentricity). **ANGER, ALIENATION AND EGOCENTRICITY** became the main devices we employed to protect us from feeling bad about ourselves while we carried out the increasingly soul-destroying search for the redeeming understanding of our corrupted condition.

[24] The problem with defensively attacking, denying and trying to prove wrong the unbearable implication that we are evil, worthless beings for having destroyed our all-loving and all-sensitive original instinctive self or soul was that while it brought us some relief we were continually adding to the levels of denial/psychosis/alienation within us.

[25] Given this ever-increasing upset, the question is, what could we do when the levels of upset within us became too unbearable and destructive and we still hadn't found the rehabilitating true, instinct vs intellect explanation of our corrupted condition?

[26] It was at this point where upset had become intolerable (which we can expect would have occurred early in our 2-million-year journey from innocence to the utterly corrupted state or condition of the human race today) that more powerful forms of artificial reinforcement of ourselves—*pseudo* therapy—had to be found. (*Real* therapy depended on finding the true, redeeming and healing instinct vs intellect biological explanation of our corrupted condition.)

Drawing by Jeremy Griffith © 2017 Fedmex Pty Ltd

[27] The first of these more powerful forms of pseudo therapy that we developed was the practice of **SELF DISCIPLINE**. As depicted above, we learned to contain and conceal the true extent of our, by now, inner fury at being unjustly condemned as evil when we intuitively didn't believe we were, and instead manufactured a calm, controlled and even a compassionate, considerate-of-others, selfless exterior. We *civilised* our upset. While we weren't able to eliminate our distressed state with understanding, we were able to restrain and hide it. Civility has disguised the volcanic levels of upset that actually exists within us humans. In order to not be overcome by the true negativity of our situation we have had to, as we say, 'put on a brave face', 'keep our chin up', 'stay positive', 'keep up appearances', pretend we are healthy and happy when we are actually the opposite,

namely incredibly psychotic and depressed. Basically, we became the absolute masters of lying and delusion, almost completely fake phonies—which, it absolutely needs to be emphasised, conversely reveals just how astronomically brave, utterly heroic and incredibly wonderful we humans actually are!!

[28] J.M.W. (William) Turner's *Fishermen at Sea*, reproduced below, captures something of the astronomical heroism of the human race struggling for 2 million years through a terrible, lonely darkness of guilt-stricken bewilderment and seeming meaninglessness, where, as the Biblical prophet Isaiah wrote, **'justice is far from us, and righteousness does not reach us. We look for light, but all is darkness; for brightness, but we walk in deep shadows. Like the blind we grope along the wall, feeling our way like men without eyes...Truth is nowhere to be found'** (Isa. 59). Yes, as the prophet of our time, and now Nobel Laureate for Literature, Bob Dylan, sang, **'How does it feel to be on your own, with no direction home, like a complete unknown'** (*Like a Rolling Stone*, 1965).

J.M.W. Turner's *Fishermen at Sea*, 1796

[29] So the extreme denial and delusion involved in the pseudo therapy of restraining and civilising our now overly upset condition helped us cope. <u>But what happened when we still hadn't found the redeeming understanding of ourselves and the levels of upset reached intolerable levels?</u> Trying to avoid the human condition (the fear that we were evil) while we were waiting to find understanding of it, was a nightmare of ever-increasing psychosis (soul death) and neurosis (mind/thinking death), until eventually we reached a point where we simply had to find our way back to at least some genuinely therapeutic connection with our instinctive true self or soul, and some genuinely therapeutic truthful thinking. <u>At that point at least a limited form of anti-denial and soul-resuscitation, or 'truth-based therapy', simply *had* to be attempted</u>.

[30] One of the first ways we developed to reconnect with soul and truth was one of the earliest forms of <u>religion</u>, which was **<u>NATURE WORSHIP OR ANIMISM</u>**—<u>religion being the strategy of putting our faith in, deferring to, and looking for comfort, reassurance and guidance from something other than our overly upset and overly soul-estranged conscious thinking egoic self</u>. Nature was a friend of our original instinctive self or soul because our instinctive self had grown up as part of the natural world, and also while nature could be brutal it wasn't psychologically upset (basically mad) like we humans have been, so by reconnecting with nature we were rebuilding at least some connection with our original all-loving and all-sensitive instinctive self or soul, and indirectly being a little bit honest about the existence of a more innocent, denial-free world.

[31] Another way that eventually developed to counter the estrangement/alienation/loneliness of our soul-oppressed situation, and this was also an earlier form of religion or deferment to something other than what our conscious thinking self was able to understand and decide it should do, was **<u>ANCESTOR WORSHIP</u>**. Having managed to survive our soul's estrangement and mind's alienated loneliness, our ancestors could be a source of great reassurance and comfort. By revering our ancestors and enshrining their memories, they could

remain a presence in our lives to inspire, look after and guide us. Our ancestors didn't connect us with our soul like nature did because they had also been psychologically upset, but their memory did help calm our upset and by so doing allow us to be a little more soulful and truthful; a little less deranged.

[32] I should point out that when upset became really extreme — which, as will be explained in the next paragraph, occurred following the development of agriculture — we could hardly live with ourselves let alone each other, and when this happened our original love for each other very often became a case of being antagonistic towards each other. (The Biblical story of the struggle between Cain and Abel is a recognition of this — see pars 906-908 of *FREEDOM*.) And the more we stopped being fond of each other, the more we stopped wanting to remember our not-so-lovable 'loved ones'. But before love died like this, we did so love each other that we didn't want to let their memory go when they died, and so in those more innocent, less upset times, adoration of our ancestors did play a very important part in our lives.

[33] What now needs to be explained is how the advent of agriculture and the domestication of animals some 11,000 years ago greatly increased the spread and growth of upset in the human race and led to the need for true religion. What was so significant about these developments in terms of escalating upset was that they caused people to live a more sedentary, less nomadic existence in close proximity and greater numbers in villages, then towns and eventually cities, and it was this greater interaction between people that dramatically increased the spread and development of upset. It was difficult enough having to cope with your own upset, let alone trying to cope with other people's upset as well. As the philosopher Jean-Paul Sartre wrote, **'Hell is other people'** (*Closed Doors*, 1944), and in a large community there are a lot of other people and thus a lot of upset everywhere, which in turn inevitably caused even more upset everywhere. It is not a relatively peaceful hunter-forager walk in the woods with innocent nature, it is a war zone of mad,

crazy, maniacal upset humans where there is little chance of our innocent soul surviving for very long. As the historian Manning Clark said, '**The bush** [wilderness] **is our source of innocence; the town is where the devil prowls around'** (*The Sydney Morning Herald*, 18 Feb. 1985). (Again, this will be discussed later in this book, but it should be noted here that the left-wing proposed that it was the agricultural revolution, and the sedentary, competition-for-possessions lifestyle it enabled, that supposedly either led to selfish instincts within us coming to dominate some selfless instincts within us, or as Critical Theory holds, just caused humans to behave selfishly, become capitalistic and so forth. However, as just explained, while the sedentary existence did dramatically increase upset, it didn't cause supposed selfish instincts to become dominant over selfless instincts as the Left argue, or simply create selfish behaviour through competition. As explained earlier, our instinctive nature is to be entirely selfless; our selfishness emerged when we became psychologically upset sufferers of the angry, egocentric and alienated human condition.)

5. As upset became unbearable, we invented Religion

[34]The next question is, what happened when, after the advent of agriculture, upset developed into an utterly unbearable state? The answer is we created the aforementioned **TRUE RELIGION**, where, in its most developed form, we deferred to and even worshipped the manifestation of our true self or soul in the form of one of the exceptionally rare individuals who had largely escaped encounter with all the upset in the world during their upbringing—exceptionally-innocent-of-upset, sound, truthful-thinking individuals we refer to as prophets.

[35]Basically, in this prophet-focused form of True Religion, when we humans became extremely upset we could decide to end our participation in humanity's heroic but upsetting battle to find

knowledge and instead place our hope and <u>faith</u> in, and live through supporting, the soundness and truth of a prophet's life and words. Rather than adhering to what our now overly upset self wanted to do and say, we could defer to, be guided by and try to emulate the soundness and truth of the prophet's life and words. This could bring immense relief to our extremely corrupted condition—as it says about Christianity in the Bible, **'if anyone is in Christ, he is a new creation; the old has gone, the new has come!'** (2 Cor. 5:17). (See F. Essay 39 for a more complete description of Christ's soundness and truthful thinking, and contribution to humanity.) Through our deferment to a religion's prophet we could be to a degree 'born again' from our corrupted condition back to a more innocent like state.

[36] <u>There were two forms of True Religion that developed before prophet-focused forms of True Religion emerged. The first involved deferring to and showing reverence for, and even worshipping, one all-pervading and all-overseeing 'God', which, as is explained in chapter 4 of *FREEDOM*, and in F. Essay 23, we can now understand is Integrative Meaning</u>—a process that is driven by the physical law known as the 'Second Path of the Second Law of Thermodynamics', or Negative Entropy, and which explains the significance of self-lessness and therefore why **'God is love'** (1 John 4:8, 16). Basically, under the influence of Negative Entropy, atoms come together to form molecules, which come together or integrate to form compounds, which integrate to form single-celled organisms, which in turn integrate to form multicellular organisms, which then integrate to form societies, and so on. That unconditionally selfless self-sacrifice for the good of the whole is the very theme of this integrative process because it maintains wholes. Selfishness is divisive and disintegrative while selfless consideration of others maintains wholes; it is integrative. As I will come back to later, the fact that the gene-based natural selection process cannot normally develop unconditional selflessness between sexually reproducing individuals—because selfless traits don't tend to reproduce—is simply a limitation of the gene-based learning system; it does not mean that selfishness is the characteristic of existence. Integrative selflessness, or love, is the

real characteristic of existence, the all-pervading and all-overseeing theme of life, which we came to personify as 'God'.

[37]This worship of one true God or **MONOTHEISM** was introduced by two exceptionally sound, denial-free thinking prophets, the Hebrew prophet Abraham, who lived around 2,000 BC, and the pharaoh Akhenaton, who reigned in Egypt from approximately 1,350 to 1,335 BC. In their extremely upset state, humans had been deriving relief for their bewildered and distressed minds by deferring to and putting their trust and faith in (even to the extent of actually worshipping) all manner of extremely superficial, false 'gods'—creating for example a god of romance, a god of war and a god of fertility, often represented by golden idols that they made appeasing sacrifices to. The worship of one true and real God or Monotheism cleared up all that extreme superficiality and brought the focus of worship onto the existence of one great all-pervading truth in the world, which, again, we can now appreciate as Integrative Meaning. Thus when our minds became so agonised from being unable to make sense of our now horrifically corrupted condition, being able to defer to and put our trust in, and even worship, the great overarching truth of one true God was a very great relief.

[38]Following the development of Monotheism's worship of one true God, the second form of True Religion developed, which was **IMPOSED DISCIPLINE** where we adhered to a set of enlightened rules or laws that enforced more selfless, cooperative and loving social behaviour through the threat of punishment. And, like Monotheism, this development was introduced by a truthful, human-condition-confronting-not-avoiding, can-still-see-upset-and-the-need-to-fix-it, sound thinking prophet (or prophets in the case of Hinduism).

[39]The outstanding example of the addition of Imposed Discipline to the worship of one true God was the exceptionally sound Hebrew prophet Moses's creation some 3,500 years ago of the Ten Commandments to accompany Abraham's teaching of the worship of one true God, the combined teachings of which are presented in the Old Testament of the Bible. Through the Ten Commandments he had etched on stone tablets, Moses brought order to the Israelite

Nation, and eventually to much of the world. Indeed, the moral code contained in those commandments became the basis of the constitutions, laws and rules that have continued to govern much of modern society and proved vital in helping rein in extreme upset.

[40] So Monotheism accompanied by Imposed Discipline proved *very* effective in restraining upset and bringing relief to the mental anguish of being extremely upset/soul-corrupted.

[41] At this point it will be helpful for what will be described next to explain how we developed the delusion that when we behaved in a cooperative and loving way we were no longer upset people. When we came up with the strategy of self-disciplining and civilising our upset we weren't doing it to delude ourselves we were upset-free, human-condition-eliminated, good people, we were only using it to restrain our upset. However, the side-effect of feeling good about ourselves when we became civilised and didn't express our upset did sow the seeds in our minds of the idea that we could use 'do good' strategies to delude ourselves that we actually were free of corruption and thus free of the agony of the human condition. In fact, this strategy of 'doing good in order to dishonestly delude ourselves that we *were* actually a human-condition-free, ideal, good person' was so superficially relieving and thus so seductive that, as will now be described, it developed into a huge industry that eventually (which is the time we now live in) threatened to destroy the human race! And what will also be revealed is that the great benefit of True Religion over all the other forms of pseudo idealism (where you pretended you were ideally behaved and free of corruption when you actually weren't) that subsequently developed, is that True Religion was by far the least dishonest and thus least dangerous form of pseudo idealism.

[42] What happened when upset inevitably increased was that the combination of Monotheism and Imposed Discipline began to prove inadequate in bringing us relief from our corrupted condition. We needed an even more powerful form of religious faith, and that was eventually supplied by the aforementioned **PROPHET-FOCUSED**

RELIGION where we deferred to, revered and worshipped the human manifestation of soulful soundness and truth, namely the worship of the exceptionally sound prophets, for example the worship of Christ in Christianity.

[43] Rather than having good behaviour forced upon you through fear of punishment, as was the case with Imposed Discipline, Prophet-Focused Religion allowed you to feel that not only were you actively participating in goodness by not allowing your upset behaviour to express itself through fear of punishment, you could delude yourself that you had actually *become* an upset-free, good, selfless, loving, ideal person—that you were 'righteous'—which provided immense relief from the guilt of being overly upset. The apostle St Paul gave what was possibly the best sales pitch for born-again, Prophet-Focused Religious life when he wrote, **'Now if the ministry that brought death, which was engraved in letters on stone** [Moses' Ten Commandments that were enforced by the threat of punishment]**, came with glory** [because they brought society back from the brink of destruction]**... fading though it was** [there was no sustaining positive in having discipline imposed on you]**, will not the ministry of the Spirit be even more glorious? If the ministry that condemns men is glorious, how much more glorious is the ministry that brings righteousness! For what was glorious has no glory now in comparison with the surpassing glory. And if what was fading away came with glory, how much greater is the glory of that which lasts!'** (Bible, 2 Cor. 3:7-11).

[44] Thus, in coping with the now raging levels of upset in humans, the first **'glorious'** improvement on destructively living out that ferocious upset was Imposed Discipline, which was enforced through fear and punishment. But since discipline provided little in the way of joy or inspiration for the mind or **'spirit'** it was hard to maintain, it didn't **'last'**, it was **'fading'**, especially in comparison to the immensely guilt-relieving, **'righteous'**, 'do good in order to make yourself feel good' way of living offered by that next **'surpassing glory'** of Prophet-Focused Religion.

[45] What now needs to be explained is why Prophet-Focused Religion was by far the least dishonest and deluded form of pseudo idealism of all the versions of the soon-to-be-described forms of pseudo idealism that developed after it.

[46] To explain this great virtue of Prophet-Focused Religion it is first necessary to emphasise the two fundamental problems with abandoning the upsetting search for knowledge. To refer back to the Adam Stork analogy, Adam could always give up the upsetting battle to find knowledge and 'fly back on course', just obey his instincts, but, while that might make him feel good, it meant that he would be failing his responsibility to keep searching for knowledge until he found the all-important relieving understanding of his corrupted condition. As I describe in F. Essay 34, there were two particular problems with giving in to this temptation to abandon the battle to find knowledge, to 'fly back on course'. They were, firstly, that you were siding against humanity's heroic battle to find knowledge, ultimately self-knowledge, understanding of the human condition; and secondly, by 'flying back on course' and supporting cooperative, loving and selfless ideal values, you were deluding yourself you were actually a virtuous, righteous, good person, even someone who didn't suffer from the upset state of the human condition.

[47] It is understandable then that when upset anger, egocentricity and alienation became so great that abandoning the battle to find knowledge could no longer be resisted, it was these two problems of siding against the human journey to find understanding, and

dishonestly deluding yourself you were a virtuous, righteous, good person, even someone free of the agony of the human condition, that those who were abandoning the battle intuitively knew they most wanted to minimise. Well, the great benefit of Prophet-Focused Religion is that it provided this relief because when you abandoned your upset way of living and instead deferred to, put your 'faith' in, and lived in support of, the sound words and life of the exceptionally sound, denial-free-thinking, truthful prophet around whom your religion was formed, you *were* minimising these two problems. (F. Essay 39 explains how, for example, the denial-free, sound words and innocent, soulful life of the great prophet Christ created the religion of Christianity.)

[48] This is because, firstly, you were indirectly still supporting the search for knowledge because the truthful words of the prophet that you were living in support of have, in fact, been the most denial-free expressions of truth and knowledge that the human race has known. In particular, most Prophet-Focused Religions acknowledge the existence of 'God', which, again, is the personification of Integrative Meaning, and the acknowledgement of this most fundamental of all truths meant you were giving the search for truth and associated knowledge the absolute best possible alignment and guidance. And there are many other guiding truths imbued in religious scriptures; the reverence humans have developed towards them is witness to that. Christianity, for example, is remarkably aligning to truth and thus supportive of our search for knowledge, which Carl Jung recognised when he wrote that '[in Christianity] **the voice of God** [truth] **can still be heard'** (W.B. Clift, *Jung and Christianity*, 1982, p.114), and that **'The Christian symbol is a living thing that carries in itself the seeds of further development'** (*The Undiscovered Self*, 1957). The value of the denial-free truth contained in Prophet-Focused Religion was also recognised by Albert Einstein when he said **'Science** [denial-infected knowledge] **without religion** [denial-free truth] **is lame, religion without science is blind'** (*Out of My Later Years*, 1950, p.26 of 286). Yes, human-condition-avoiding mechanistic science has been so dangerously devoid of truth (see

Video/F. Essay 14 & F. Essay 40) that the only place where any denial-free truth has been maintained is in religious scripture.

[49] Secondly, by taking up support of a Prophet-Focused Religion you were minimising the problem of the dishonesty of deluding yourself you were a good, even a human-condition-free person, because of the many descriptions in religious scripture that recognise how corrupted, 'sinful', 'guilty', 'banished'-from-the-Garden-of-Eden-state-of-original-innocence we upset humans have been. Also, by acknowledging the soundness of the prophet by your worship of him, you were, by inference, acknowledging your own lack of soundness. You were indirectly being honest about your extremely upset, corrupted condition.

[50] So Prophet-Focused Religions have been a marvellously effective way for humans to indirectly support the search for knowledge, and to be honest about our corrupted condition. (See F. Essay 39 for more analysis of the benefits of religion.)

[51] The problem, however, that emerged with Prophet-Focused Religion as humans became even more upset was that the great benefit of their honesty (about our corrupted, innocence-destroyed, 'guilty', 'sinful', Integrative-Meaning-or-God-defying lives) became its liability. Basically, without the defence of the true instinct vs intellect explanation for the upset state of the human condition, the more upset we humans became, the more unbearably confronting and condemning it was having our upset, soul-corrupted state acknowledged. And so, when upset became extreme, the honesty in Prophet-Focused Religion about our corrupted state became intolerable; as the author Mary McCarthy once wrote, **'Only people who are very good can afford to become religious; with all the others it makes them worse'** (*Memories of a Catholic Girlhood*, 1957; Lloyd Reinhardt's rendition, *The Sydney Morning Herald*, 18 Jan. 1995).

[52] It was at this point that less honest, more guilt-stripped, less confronting and condemning, more delusional feel-good ways of abandoning the battle had to be found. One of the first ways of

achieving this was to modify Prophet-Focused Religions so they were less confronting. This was done by interpreting their teaching and scripture in very superficial, simplistic, literal and fundamentalist ways, and by focusing only on the relief your faith was bringing you through emotional, euphoric, 'evangelical', 'charismatic' forms of worship. Sir Laurens van der Post pointed out how dangerously 'starved and empty' of its truthfulness, guilt-stripped Christianity has become when he wrote that 'Yet the churches continue to exhort man without any knowledge of what is the soul of modern man and how starved and empty it has become...They behave as if a repetition of the message of the Cross and a reiteration of the miracles and parables of Christ is enough. Yet, if they took Christ's message seriously, they would not ignore the empiric material and testimony of the nature of the [cooperative and loving] soul and its experience of God [Integrative Meaning]' (*Jung and the Story of Our Time*, 1976, p.232 of 275).

[53] As upset further increased, atheism or disbelief in God, and non-religious secularism, became popular—to the point where the scientist Richard Dawkins has angrily said, "'Faith is one of the world's great evils, comparable to the smallpox virus, but harder to eradicate. The

whole subject of God is a bore"…those who teach religion to small children are guilty of "child abuse"' ('The Final Blow to God', *The Spectator*, 20 Feb. 1999). (Read more about science's aversion to the truth contained in religion in F. Essay 40.)

Richard Dawkins's dismissal of religion's contribution
to humanity's journey to find understanding

6. The progression over the last 200 years when upset became completely overwhelming

[54] To now describe the progression that has taken place over the last 200 years when upset has become overwhelmingly extreme and, as a result, such extreme states of dishonesty, denial and delusion have developed that the human race is facing terminal alienation and extinction.

[55] Since any form of Prophet-Focused Religion was becoming unbearable, a much more dishonest, guilt-free form of pseudo ideal-ism had to be found. The first solution was atheistic **COMMUNISM or SOCIALISM** which did not contain any recognition of the sound,

relatively upset-free, innocent, soulful world of prophets, nor acknowledgement of the condemning truth of Integrative Meaning in the form of 'God'. Instead, communism dogmatically demanded an idealistic social or communal world and denied the depressing notion of God and associated guilt.

[56]The dishonesty and delusion of communism/socialism was made very clear when its creator Karl Marx asserted that **'The philosophers have only interpreted the world in various ways; the point is** [not to understand the world but] **to change it** [just make it cooperative/social/communal]' *(Theses on Feuerbach, 1845).* No, the whole **'point'** and responsibility of being a conscious being *is* to understand our world and our place in it—ultimately, to find understanding of our seemingly horribly flawed, angry, egocentric and alienated human condition. While more will be said about Marxism shortly, the fundamental problem with communistic and socialistic Marxism was that its dogmatic, **'just change it'** imposition of cooperative and selfless 'politically correct' idealism denied humans the freedom to be competitive, aggressive and selfish while they carried out their all-important upsetting search for knowledge, ultimately for self-knowledge, understanding of our corrupted condition. Dogma wasn't the cure for our corrupted condition, it was the oppressive poison.

[57] While this oppression of freedom was why socialism/communism was initially largely decried as a political system for managing society, at the personal level, <u>the limitation of socialism/communism was that while there was no confronting innocent prophet present, there was an obvious acknowledgment of the condemning cooperative ideals</u>. So in time, as levels of insecurity rose so too did the need for an even more evasive and dishonest guilt-free form of pseudo idealism to live through. This was supplied by the **NEW AGE MOVEMENT** (the forerunners of which were the Age of Aquarius and Peace Movements) in which all the negatives of humans' massively corrupted condition were transcended in favour of taking up a completely escapist, think-positive, human-potential-stressing, self-affirming, motivational, feel-good-about-yourself approach.

[58] In talking about how he became **'a personal growth junkie'**, the comedian Anthony Ackroyd summed up the extremely deluded artificiality of the New Age Movement when he said: **'What are millions of us around the globe searching for in books, tapes, seminars, workshops and speaking events? Information to enhance our lifestyles and**

enrich our experience on this planet? Certainly...But I smell something else in the ether. Something more desperate and deluded. A worrying snake-oil factor that is spinning out of control. It's the promise of salvation. Salvation from the basic rules of human life. This is the neurotic aspect of the human potential movement. This hunger for a get-out-of-the-human-condition-free card' (*Good Weekend*, *The Sydney Morning Herald*, 13 Sep. 1997). Yes, to 'get out of the human condition' we had to confront and solve it, not deny and escape it; our 'desperate and deluded' attempts to escape it only made it worse. As the philosopher Thomas Nagel recognised, 'The capacity for transcendence brings with it a liability to alienation, and the wish to escape this condition...can lead to even greater absurdity' (*The View From Nowhere*, 1986, p.214 of 256).

[59]The limitation of the New Age Movement was that while it didn't remind humans of the cooperative ideals, the focus still remained on the issue of humans' variously troubled, alienated, upset, innocence-destroyed state. So the next level of delusion dispensed with the problem of alienation by simply denying its existence. The **FEMINIST MOVEMENT** maintains that there is no difference between people, especially not between men and women. In particular it denied the legitimacy of the egocentric male dimension to life. As is explained in paragraphs 770-777 of *FREEDOM*, far from being destructive villains, men turn out to be nothing less than the heroes of the story of life on Earth!

[60] Being based on extreme dogma, the feminist movement could not and has not produced any real reconciliation between men and women, rather, as this quote points out, **'What happened was that the so-called Battle of the Sexes became a contest in which only one side turned up. Men listened, in many cases sympathetically but, by the millions, were turned off'** (Don Peterson, review of *The Myth of Male Power* by Dr Warren Farrell, *The Courier Mail*, Jun. 1994). (Read about how the explanation of the human condition brings an end to the so-called 'war of the sexes' in F. Essay 26.)

[61] The limitation of feminism was that while it sought to dispense with the problem of humans' divisive reality, humans were *still* the focus of attention and that was confronting. So again, as upset increased and an even more evasive and dishonest form of pseudo idealism to support was needed, the **ENVIRONMENT or GREEN or CLIMATISM MOVEMENT** emerged in which there was no need to confront and think about the human state since its focus was away from self and onto the environment.

[62] In response to a question about people's attachment to climatism (focus on climate change), the psychologist Jordan Peterson spoke the truth when he said, **'People have things within their personal purview** [range of experience] **that are difficult to deal with and that they're avoiding and generally the way they avoid them is by adopting pseudo-moralistic stances on large-scale social issues, so they look good to their friends and neighbours'** (Australia's ABC program *Q&A*, 25 Feb. 2019). The inherent dishonesty and irresponsibility of this movement was summed up by these quotes: **'The trees aren't the problem. The problem is us'** (*Simply Living* mag. Sep. 1989), and **'We need to do something about the environmental damage in our heads'** (*Time*, 24 May 1993). The bumper sticker **'Save the Humans'** that parodies the green movement's **'Save the Planet'** slogan also makes the point about how evasive of the real issue the environmental movement is—as does this statement: **'Environmentalism has largely superseded Christianity as the religion of the upper classes in Europe and to a lesser extent in the United States. It is a form of religious belief which fosters a sense of moral superiority in the believer, but which places no importance on telling the truth** [about the real issue of our corrupted condition]' (Ray Evans, *Nine Facts About Climate Change*, 2006).

[63] So while it has been said that **'The environment became the last best cause, the ultimate guilt-free issue'** (*Time*, 31 Dec. 1990), the Environment Movement has an undermining limitation in that it still contained a condemning moral component. If we were not responsible with the environment, 'good', we were behaving immorally, 'bad'. In addition, the purity or innocence of nature contrasted with humans' lack of it. Clearly, as upset further increased, an even more evasive, human-condition-avoiding, **'guilt-free'** form of idealism was going to be needed, but before presenting that development, a description needs to be given of how rapidly upset was growing.

7. The crescendo of upset that has developed in the last 50 years

[64] At this point the march of ever increasing upset was knocking on the door of terminal alienation, which the following graph that I presented in *TI* illustrates.

The race between self-destruction and self-understanding

[65] Indeed, younger generations are now so alienated and in need of blocking out any confronting truth about the corrupted state of the human condition that any deeper thinking at all has become impossible—which is why universities provide 'trigger warnings' on any confronting subject matter, 'safe spaces' for students to escape such confrontations, and protection from 'microaggressions' that en- counters with any truth about the extent of the upset in themselves or in the whole human race represent to them. A 2017 article in *The Australian* newspaper evidences this trend, with journalist Jennifer Oriel reporting that 'half of undergraduate students [now] think it is accept- able to silence speech they feel is upsetting' (25 Sep. 2017). In fact, Millennials (also known as the 'Y-generation', those born between 1982 and

1998) and subsequent 'Z' and 'Alpha' generations are described as **'snowflakes'** because they melt if placed under any pressure.

2017 cover picture to article titled 'Professor Encourages Students To Choose Their Own Grades To "Reduce Stress"' (Joshua Caplan, *The Gateway Pundit* political website, 7 Aug. 2017)

Universities have been places of learning, but how can you learn and seek truth if you aren't prepared, expected or even challenged to think anymore?

[66] Unaware of the instinct vs intellect explanation of the human condition and the ever-increasing levels of anger, egocentricity and

alienation that the struggle to find that explanation has unavoidably caused, researchers have sought to blame the on-rush of alienation/ block-out/denial/soul-estrangement/psychosis in emerging genera- tions on such superficial causes as the overuse of communication technology, such as the internet, social media, smartphones and computer games. While this technology has greatly increased and spread alienation by exposing children to all the upset in the world which destroys their innocent, happy, trusting and loving souls, its overuse, even addiction, is the result of a massive need for distraction from overwhelming internal psychological pain.

[67] As the psychologist Arthur Janov pointed out: 'The brain [of young people today] is busy, busy, dealing with the pain'; with the result that 'when there is stimulation from the outside…it meets with a very active brain which says "Whoa there. Stop the input. I have too much going on inside to listen to what you ask for"…Of course, the kid is agitated out of his mind, driven by agony inside. We want her to focus on 18th century art and she is drowning in misery' ('Once More on Attention Deficit Disorder', 4 Apr. 2013).

[68] Indeed, almost 10 years ago now in 2013 an art teacher at one of Sydney's leading private schools told me that 'while only two years ago students were able to sit through a half hour art documentary, I now know I lose them after only eight minutes; today's students' attention span is that brief!' This comment mirrors an observation made by the political scientist David Runciman in a 2010 BBC documentary series about the internet: 'What I notice about students from the first day I see them when they arrive at university is that they ask nervously "What do we have to read?" And when they are told the first thing they have to read is a book they all now groan, which they didn't use to do five or ten years ago, and you say, "Why are you groaning?", and they say "It's a book, how long is it?"' (*Virtual Revolution*, episode 'Homo Interneticus'). The same documentary also included the following insightful statement from Nick Carr, the author of *Is Google Making Us Stupid?*: 'I think science shows us that our brain wants to be distracted and what the web does by bombarding us with stimuli and information it really plays to that aspect of our brain, it keeps our brain hopping and jumping and unable to concentrate.'

[69] In the case of social media such as Facebook, Instagram, Twitter and TikTok, it allows people to be preoccupied/distracted (from the human condition) all day long with inane, frivolous, narcissistic, superficial self-promotion and gossip. The result of this extreme distraction from the **'agony inside'** is that **'The youth of today are living their lives one mile wide and one inch deep'** (Kelsey Munro, 'Youth skim surface of life with constant use of social media', *The Sydney Morning Herald*, 20 Apr. 2013). Yes, **'the net delivers this shallow, scattered mindset with a vengeance'** ('The effects of the internet: Fast forward', *The Economist*, 24 Jun. 2010). As one member of the Millennial generation self-analysed, **'Alone and adrift in what [Professor] de Zengotita calls our "psychic saunas" of superficial sensory stimulation, members of my generation lock and load our custom iTunes playlists, craft our Facebook profiles to self-satisfied perfection, and, armed with our gleefully ironic irreverence, bravely venture forth into life within glossy, opaque bubbles that reflect ourselves back to ourselves and safely protect us from jarring intrusions from the greater world beyond'** (Tom Huston, 'The Dumbest Generation? Grappling with Gen Y's peculiar blend of narcissism and idealism', *EnlightenNext*, Dec. 2008-Feb 2009).

[70] What all this exhaustion of soul has finally led to in the last few years is that young people are starting to give up on having anything to do with anything. A 2018 study by **'social scientists seek[ing] explanations for millennials' moderate ways'** said that **'something is up'**, **'teenagers seem lonelier than in the past'**, and talked of **'young people becom[ing] virtual hermits'**, and of **'taking it slow. They are slow to drink, have sex and earn money.'** It said that **'teenagers are getting drunk** [on mind-numbing alcohol] **less often'** and **'other** [escapist] **drugs are also falling in favour'**, and **'young people [are] harming each other much less than they used to. Fighting among 13 and 15-year-olds is down across Europe'**, and **'teenagers are also having less sex'**. It said that **'In short, young people are less hedonistic and break fewer rules than in the past. They are "kind of boring"'** ('Teenagers are better behaved and less hedonistic nowadays but they are also lonelier and more isolated', *The Economist*, 10 Jan. 2018). The truth is that the last refuge for the terminally alienated is dissociation from the world. In fact, the epidemic numbers of children now suffering from the extremely agitated mental

condition of ADHD, and the completely-dissociated-from-the-world
state of autism, shows how this terminal stage of upset is upon us.

[71] The following stories in the media provide a powerful snapshot
of the rapidly increasing levels of alienation in society over the last
20 years.

<u>2003</u> *TIME* cover story about **'A medicated generation'** of children **'suffering
from' 'exploding rates'** of **'the alternatively depressive and manic mood
swings of bipolar disorder (BPD)'** and **'attention-deficit/ hyperactivity
disorder (ADHD)'**, with experts admitting **'we don't know exactly why the
incidence of psychopathology is increasing in children and adolescents'**
(8 Dec. 2003). Yes, living in denial of the human condition meant we were in no
position to understand the effects of the human condition.

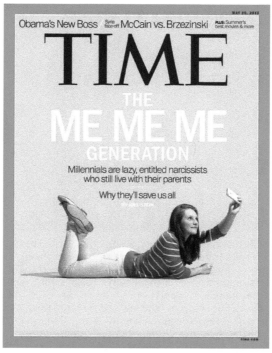

2013 *TIME* magazine cover story about the **'narcissism epidemic'** of the **'80 million strong...biggest age group in American history' 'Me Me Me Generation'** of **'Millennials'**. The article said that this **'narcissism',** which is a **'personality disorder'** where **'people try to boost** [low] **self-esteem',** may mean that the Millennials' **'self-centredness could bring about the end of civilization as we know it'.** Saying that **'Millennials'** will **'save us all'** was a reverse-of-the-truth lie designed to put a positive spin on this real sign of **'the end of civilization'** (20 May 2013).

This **2016** *TIME* magazine cover story, headed **'The Kids Are Not All Right: American teens are anxious, depressed and overwhelmed. Experts are struggling over how to help them'**. Describing a **'spectrum of angst that plagues 21st-century teens'**, like **'anxiety'**, **'depression'**, **'self-loathing'**, **'fragility'**, **'sadness'** and **'hopelessness'**, the article begins and ends with harrowing descriptions of teens **'cutting'** and **'hurting themselves'** because **'physical pain may relieve the psychological pain'**. **'Self-harm'**, it says, **'does appear to be the signature symptom of this generation's mental health difficulties'** (7 Nov. 2016).

A **2017** article in *The Economist* reported that **'the suicide rate for** [American] **15 to 19-year-olds shot up between 2007 and 2015, increasing by 31 per cent for boys, and more than doubling for girls'** (23 Nov. 2017).

A **2017** article reported that Australia's ***'Five Year Mental Health Youth Report…* found the proportion of young people likely to have serious mental illness rose from 18.7 per cent in 2012 to 22.8 per cent in 2016'**, and went on to say that **'There is a tidal wave of mental health issues in the schools'**
(*Sydney Morning Herald*, 19 Apr. 2017).

[72]The fact is the paralysed, can't-cope-with-anything, aptly titled **'snowflake'** millennials are now describing *themselves* as **'The Burnout Generation'**—although, unable to acknowledge their extreme psychosis/soul-death/alienation from the end play state of humanity's heroic search for the explanation of why we corrupted our cooperative and loving instinctive self or soul as being the real reason for their burn-out, they are blaming it on such superficial causes as stress from the overuse of **'smartphones'** and from **'the 2008 financial crisis'**; to being **'scared'** of the world due to **'intensive'** **'helicopter parents'**; and in general to the **'mental load'** produced by the **'systems of capitalism and patriarchy'**! (Anne Helen Petersen, 'How Millennials Became The Burnout Generation', BuzzFeednews.com, 5 Jan. 2019.) As is explained elsewhere in this book, materialistic **'capitalism'** and the **'patriarch**[al]**'** work of solving the human condition are what has saved the human race from extinction, but such is the despair and madness of thinking in the human mind now that our saviours are portrayed as the villains.

[73]In terms of understanding the non-superficial, *real* reason for the increase in upset from one generation to the next, which was the upsetting effects of humanity's heroic search for knowledge, there were two components. Firstly, there was the upset each person developed from their own experiments in self-management when there was no understanding of how such experiments could lead to outcomes that weren't consistent with what our cooperative and loving instinctive self or soul expected—which, without that understanding, were coped with by defensively attacking, denying and trying to prove wrong those implied criticisms of our experiments; which, again, is the upset that self-management caused while we didn't understand that we weren't bad to search for knowledge and make mistakes. The second, and by far the greatest reason for the increase in upset that has occurred from generation to generation is the soul-destroying influence of the existing levels of upset in your society from all the upsetting searching for knowledge that has taken place before you arrived in the world. Each generation is born expecting to encounter the all-loving and all-sensitive world

we humans originally lived in, and the more upset/soul-corrupted the world we actually encounter is, the more bewildered, hurt and damaged our soul became by that encounter. When the great psychiatrist R.D. Laing wrote that **'To adapt to this world the child abdicates its ecstasy'** (*The Politics of Experience* and *The Bird of Paradise*, 1967, p.118 of 156) he was recognising how sensitive our soul is and how much it dies when it doesn't encounter the love and happiness it expects. <u>Admitting the loving sensitive nature of our instinctive self or soul is what allows us to understand what is crippling younger generations. What the internet does then that is so destructive of young people is expose them to no end of soul-terrifying-and-deadening trauma.</u> Playwright Samuel Beckett was only slightly exaggerating the brevity today of a truly loved, soulful, happy, innocent, secure, nurtured-with-unconditional-love-from-mothers-and-reinforced-with-unconditional-kindness-from-fathers, sane, trauma-free life when he wrote, **'They give birth astride of a grave, the light gleams an instant, then it's night once more'**! (*Waiting for Godot*, 1955.)

Grotesque as it is, this detail from Goya's 1819 painting of Cronus devouring his child is a fair representation of how destructive our upset is of children's innocence.

[74] And the COVID-19 pandemic that began in 2020 has so magnified the already unbearable levels of upset in younger generations that our youth are now living in a *completely* overwhelmed and traumatised state. If, since 2 million years ago, human life had not been based on living in denial of the truth of our corrupted condition while we couldn't explain it, and we could see how terminally psychologically upset and soul-corrupted we humans have become, every publication in the world would carry the headline 'We have to solve the human condition right now or all is lost.' The following famous painting by Théodore Géricault captures the true extent of the exhaustion of the human race and the now absolutely desperate need to find the relieving understanding of the human condition.

The Raft of the Medusa by Théodore Géricault, 1818-19

[75] Unfortunately, because we are living in alienated denial of the truth of our soul-corrupted condition, apart from some superficial recognition of our psychologically distressed state, such as the *TIME* magazine cover stories, there is not even any mention of the linchpin issue of the human condition in the media! Truly, as that breathtakingly

honest psychiatrist R.D. Laing also wrote, 'Our alienation goes to the roots [p.12 of 156] ...the *ordinary* person is a shrivelled, desiccated fragment of what a person can be...we hardly know of the existence of the inner world [p.22] ...The condition of alienation, of being asleep, of being unconscious, of being out of one's mind, is the condition of the normal man [p.24] ...between *us* and It [our true selves or soul] there is a veil which is more like fifty feet of solid concrete [p.118] ...The outer divorced from any illumination from the inner is in a state of darkness. We are in an age of darkness [p.116] (*The Politics of Experience* and *The Bird of Paradise*, 1967). 'We are dead, but think we are alive. We are asleep, but think we are awake...We are *so* ill that we no longer feel ill, as in many terminal illnesses. We are mad, but have no insight [into the fact of our madness]' (*Self and Others*, 1961, p.38 of 192). 'We are so out of touch with this realm [this issue of the human condition] that many people can now argue seriously that it does not exist' (*The Politics of Experience* and *The Bird of Paradise*, p.105). Yes, the level of madness, and conversely the level of bravery of the human race, is astronomical! Even though not many are aware of it yet, MERCIFULLY THE HUMAN CONDITION HAS BEEN SOLVED—THE INSTINCT VS INTELLECT, RECONCILING, REDEEMING AND REHABILITATING BIOLOGICAL EXPLANATION FOR WHY WE CORRUPTED OUR SPECIES' ALL-LOVING AND ALL-SENSITIVE BUT NON-CONSCIOUS ORIGINAL INSTINCTIVE SELF OR SOUL HAS BEEN FOUND! These clearly un-resigned, denial-free, honest lyrics from the 2010 *Grievances* album of the young American heavy metal band With Life In Mind also powerfully reinforce how 'desperate for the answers' about the human condition we have been, and also how 'Fear is driven into our [young people's] minds everywhere we [they] look', and how 'scared to use our minds' the human race has been: 'It scares me to death to think of what I have become...I feel so lost in this world', 'Our innocence is lost', 'I scream to the sky but my words get lost along the way. I can't express all the hate that's led me here and all the filth that swallows us whole. I don't want to be part of all this insanity. Famine and death. Pestilence and war. [Famine, death, pestilence and war are traditional interpretations of the 'Four Horsemen of the Apocalypse' described in Revelation 6 in the Bible. Christ referred to similar 'Signs of the End of the Age' (Matt. 24:6-8 and Luke 21:10-11).] A world shrouded in darkness...

<u>Fear is driven into our minds everywhere we look</u>', 'Trying so hard for a life with such little purpose...Lost in oblivion', 'Everything you've been told has been a lie...We've all been asleep since the beginning of time. <u>Why are we so scared to use our minds?</u>', 'Keep pretending; soon enough things will crumble to the ground...If they could only see the truth they would coil in disgust', 'How do we save ourselves from this misery...<u>So desperate for the answers</u>...We're straining on the last bit of hope we have left. No one hears our cries. And no one sees us screaming', 'This is the end.'

[76] The completely exhausted inclination in emerging generations to give up on life is going to have, indeed is already having, very serious political and thus social consequences. Basically, rather than continue the heroic but psychologically upsetting, anger, egocentricity and alienation-producing battle against our instincts that unavoidably results from searching for knowledge, ultimately for self-knowledge, the all-important, psychologically relieving understanding of our corrupted human condition, they want to quit the battle, throw in the towel, give up on finding knowledge. Exhausted by the battle, they are throwing their hands up and saying 'Let's just stop this struggle and be good to each other'—vote for outrageously irresponsible, extreme pseudo idealistic socialists like Jeremy Corbyn in British politics and Bernie Sanders in US politics. To quote from a 2016 article in *The Atlantic*, '**And if there's one thing people are learning about this young generation, it's that they are liberal. Even leftist. Flirting with socialist. In Iowa, New Hampshire, and Nevada, more than 80 percent of voters under 30 years old voted for Bernie Sanders, a democratic socialist so outside the mainstream of his party that he's not even a member**' (Derek Thompson, 'The Liberal Millennial Revolution', 1 Mar. 2016).

[77] As will be described shortly when the development of Postmodernism and then Critical Theory is explained, this abandonment of the immensely upsetting, heroic battle the human race has been involved in to find understanding of the human condition has developed into not just a totally dishonest attitude, but into totally dishonest philosophical theories about the very nature and purpose of human existence!

[78] Yes, it certainly is endgame for the human race when humans completely give up on thinking, which is essentially what has happened to those born since 1982 (those under the age of 39 in 2021 when this book was written). Indeed, a further indication of how terminally alienated the human race has become is that we get virtually no response to our online advertisements that promote understanding of the human condition from those now under the age of 39. The all-important issue of our species' troubled human condition has become just too unbearable to even begin to think about for all but a rare few who are younger than that. Some of us from the Sydney WTM who attended a seminar on online marketing in 2017 were actually told that **'conventional advertising doesn't work with anyone under the specific age of 35'**, and that they require **'much more backed-off, minimalist, even oblique messaging'** (Salesforce World Tour). Business is finding it hard to find functional Millennials to employ. The chair of Australia's national broadcaster, Ita Buttrose, complained that Millennials **'so lack resilience they need hugging'** (*Sydney Morning Herald*, 23 Jul. 2020). They are preferring to work from home, want to take their dog to work, ban phone calls from management because they find them too confrontational, want the business to have a climate agenda and be 'woke' so they can feel good about themselves. Elon Musk was so frustrated with their paralysis at his Tesla company he suggested they should go and **'pretend to work somewhere else'** — an inability to work that leads to society's welfare and disability funds being drained! The snowflake frailty of Millennials and subsequent generations, the complete shutdown of the human mind that is taking place, the arrival of terminal alienation, is *very* real! In fact, with regard to our marketing, we have learnt that advertising to under 39s (in 2021) is so futile that we have no choice but to avoid it — which means that the initial appreciation and support of the human-race-saving understanding of the human condition is going to have to come from people over 39, which is a stark measure of the extremely serious, fast-running-out-of-time situation the human race is in.

[79] It should be emphasised that what is said about Millennials and subsequent generations is not an attack on them. As is explained here and in *TI*, the unavoidable, heroic price the human race had to be prepared to pay for searching for knowledge was ever increasing levels of upset anger, egocentricity and alienation, which has led to the extremely upset Millennial and subsequent generations. So a generation's particular level of upset is not their fault, rather it is a consequence of where it happens to fall in humanity's progression of upset. Read more about the march of upset to terminal levels in F. Essay 55, and chapters 8:16A-Q of *FREEDOM*.

8. The development of truthless Postmodern Deconstructionism and Political Correctness

[80] So while the levels of alienation among those born after 1982 had become so extreme that any engagement with deeper meaningful thinking had become virtually impossible, those born in the decades leading up to that time were also approaching these extreme levels of alienation, and being so alienated and insecure about their extreme level of upset, a form of pure 'idealism' had to be developed for those born during those years and beyond where *any* confrontation with the, by now, extremely confronting and depressing moral dilemma of the human condition was totally avoided. This need for a totally guiltless, non-confronting form of deluded 'idealism' that enables a person to feel they are doing good and are an upset-free, 'I don't suffer from the human condition', good person was met by the development of the **POLITICALLY CORRECT MOVEMENT** and its intellectual equivalent, the **POSTMODERN DECONSTRUCTIONIST MOVEMENT**. These were pure forms of pseudo idealistic dogma that fabricated, demanded and imposed equality in complete denial of the reality of the underlying issue of the reasons for the different

levels of alienation between individuals, sexes, ages, generations, races and cultures. For example, the politically correct argue that the children's nursery rhyme *Baa Baa Black Sheep* is racist and must instead be recited as **'Baa Baa Rainbow Sheep'**! (*The Telegraph*, 24 Jan. 2008.)

[81] Postmodernism has been described by a journalist as **'a bewilderingly complex school of continental philosophy, or pseudo-philosophy'** of **'intellectual assumptions—[that] truth is a matter of opinion, there is no real world outside of language and hence no facts independent of our descriptions of them'** (Luke Slattery, *The Australian*, 23 Jul. 2005). While language is artificial it nevertheless models a real world, so to say that just because language is artificial there can be no universal truths is ridiculous, but when the need to escape the truth becomes desperate, any excuse will do; just baffle the world, and yourself, with intellectual baloney. In his 2001 book, *The Liar's Tale: A History of Falsehood*, Jeremy Campbell described **'postmodern theory'** as having elevated **'lying to the status of an art and neutralised untruth'**. It **'neutralised untruth'** because by denying the existence of the whole issue of humans' variously upset state it made any discussion of such differences impossible.

[82] In his insightfully titled 1987 book, *The Closing of the American Mind*, the political scientist Allan Bloom wrote of the devastating effects of teaching the extremely dishonest, pseudo idealistic post-modern, deconstructionist thinking in universities—as summarised in this book review: **'we are producing a race of moral illiterates, who have never asked the great questions of good and evil, or truth and beauty, who have indeed no idea that such questions even could be asked...As Mr Bloom says..."deprived of literary guidance they [students] no longer have any image of a perfect soul, and hence do not long to have one. They do not even imagine that there is such a thing"...If the classics are studied at all in the universities they are studied as curiosities in the humanities departments, not as vital centres of the liberal tradition, and not as texts offering profound insight into the human condition'** (Greg Sheridan, 'The Closing of Our Minds', *The Australian*, 25 Jul. 1987). Of course, the whole point of the postmodern, politically correct culture was to avoid **'the great questions'** about our species' all-loving and all-sensitive, original instinctive self or **'soul'** and what has happened to it, namely the question of our self-corruption and resulting denial/alienation/psychosis, the issue of **'good and evil'**, **'the human condition'**, **'truth'**.

[83] Yes, the determined denial of humans' variously immensely corrupted condition in the politically correct, postmodern, deconstruction movements meant that instead of persevering with humanity's heroic search for knowledge and finding the true instinct vs intellect explanation of the human condition that actually reconciles and thus 'deconstructs' the good versus evil dialectic and, by so doing, takes humanity beyond or 'post' the existing upset, 'modern', human-condition-denying, dishonest, unreconciled world to a human-condition-understood-and-ameliorated, upset-free, 'correct' one where everyone lives cooperatively and lovingly, as these movements in effect claimed they were doing, they were leading humanity further away from that solution and ideal state.

[84] For immensely upset humans, however, the limitation of politically correct, postmodern deconstructionist movements was that the focus was very much on relieving yourself of the guilt of your corrupted condition by, in the extreme, eliminating the whole idea of truth. What was needed was a program that focused more on the progression of the whole human race towards a more ideal state—a feel-good, virtue-signalling philosophy that contrived a way (because it couldn't be achieved honestly by facing the truth of our corrupted condition and by so doing actually solving that condition with redeeming understanding) to transport the human race from its competitive and aggressive state to a more just and 'equitable' world where *all* humans lived cooperatively, selflessly and lovingly.

[85] Basically, because the upsetting battle to find understanding of ourselves had all become too unbearable, there was a shift from just wanting personal guilt-free relief to wanting to create a whole new world; to 'reset' the whole foundations of society. 'Enough suffering, let's just get out of the horror world we live in and create a new happy and loving world, let's fake it to make it' was the thinking. As we will see, while Marxism had largely been avoided because we intuitively knew it oppressed the freedom humans needed to

search for knowledge, this great need for a philosophy that took the whole human race to a more cooperative and loving state led to the resurrection of Marxism's dogmatic imposition of cooperative and loving behaviour.

[86]What was actually subconsciously 'decided' was that the upset in yourself and in the world had become so great that, even if it was fraudulently achieved, the dream that all humans have held deep within them that one day a new world free of our competitive and aggressive human condition would be possible simply had to be implemented. That is essentially what happened—it was subconsciously decided that the great goal of the whole human journey of conscious thought and enquiry of bringing an end to the corrupted state of the human condition and our return to idyllic Eden had to be faked. But, as we will see, what was manufactured was *so* utterly dishonest and deluded it was taking humanity straight to extinction! This fraudulent, outrageously dishonest, incredibly dangerous contrivance is **CRITICAL THEORY**—with its offshoots of Critical Race Theory (CRT) and Critical Gender Theory, and their manifestations of 'Identity Politics', 'Woke' ideology, 'Cancel Culture', and the 'Great Reset' of society.

[87]Before explaining Critical Theory, the progression of the biological arguments that eventually led to the resurgence of Marxism needs to be described.

9. The progression of biological arguments that led to the rise of Critical Theory

[88]At the beginning of my booklet *Transform Your Life And Save The World* (*TYL*) (which is a summary of *FREEDOM*), I quote Sir Bob Geldof saying to me, **'We're not all going to turn into people who are all hugging each other Jeremy because we're all competitive by nature. The question is how do we relieve ourselves from these unchangeable competitive, selfish and aggressive primal instincts in us?'** As I go on to explain in *TYL*,

Sir Bob's assertion that we're not going to start being cooperative and loving and **'all hugging each other'** **'because we're all competitive by nature'**, shows that he, like almost everyone has been doing, is using the false 'savage instincts' excuse to explain our divisive behaviour. And when Sir Bob added that **'The question is how do we relieve ourselves from these unchangeable competitive, selfish and aggressive primal instincts in us?'**, he was defending his well-known advocacy of the left-wing position—that is now rapidly being replaced by Marxist-based Critical Theory—which argued that along with some selfless instincts we supposedly also have savage, selfish, competitive and aggressive instincts. And since we are born with these supposed savage instincts, we can't change them, and therefore, wherever they overly assert themselves, which the Left see as happening everywhere, the Left claim we have no choice but to dogmatically impose cooperative and loving ideal values on those supposed **'unchangeable competitive, selfish and aggressive primal instincts'**.

[89] Certainly, if it *was* true that we have savage instincts then the logic of Sir Bob's left-wing culture where cooperative and loving values have to be dogmatically imposed *would* be justified. BUT, as I explained in *TI*, we don't have savage competitive and aggressive instincts, rather we have unconditionally selfless, cooperative and loving moral instincts, which means the fundamental premise of this left-wing philosophy is wrong.

[90] To explain more about Sir Bob's assertion that along with selfless instincts we also have **'unchangeable competitive, selfish and aggressive primal instincts'**: as I mentioned earlier when summarising *TI*'s description of our condition, to counter the selfishness-justifying, individualistic, right-wing assertions that we humans are naturally selfish because we supposedly have savage, must-reproduce-our-genes, 'survival of the fittest' instincts, the Left set about developing selflessness-emphasising, communalistic theories that maintain that while we do have some savage, competitive and aggressive, must-reproduce-our-genes instincts we also have some selfless instincts derived from situations where supposedly cooperation

was a more successful survival strategy than competition. After presenting a history of selfishness-justifying right-wing dishonest biology in chapters 2:8 to 2:11 of *FREEDOM*, in chapter 6:9 I present a history of this left-wing dishonest biology which argues that a group who are selfless and cooperative will defeat a group who are selfish and competitive, and that is supposedly how we developed some selfless, cooperative instincts—which, as I have pointed out before, is actually biologically impossible because of **'the tendency of each group to quickly lose its altruism through natural selection favoring cheaters** [selfish, opportunistic individuals]**'**, as the biologist Jerry Coyne pointed out. In that chapter 6:9 I describe, for example, how in the 1960s the behaviourist Konrad Lorenz wrote of *behaviour* having **'a species-preserving function'** (*On Aggression*, 1963). I then describe how when right-wing biologists pointed out the falsity of this 'group selection' theory—such as George Williams in his 1966 book, *Adaptation and Natural Selection*—the left-wing then tried to maintain that we do have unconditionally selfless moral instincts by arguing that they are derived from by-products of natural selection. An example of this 'pluralistic' approach was that put forward by biologists Stephen Jay Gould and Richard Lewontin who in 1979 described it as **'a lot of [building] cranes'** acting in conjunction with **'natural selection'** ('Darwin Fundamentalism', *The New York Review of Books*, 12 Jun. 1997, in which Gould elaborated upon his and Lewontin's by-products or 'Spandrels' theory). But unable to identify precisely what these **'by-products'/'spandrels'/'cranes'** were, the Left were again forced to retreat to the now highly discredited 'cooperation is more advantageous than competition and can therefore be selected for', group-selection-type argument. And so, in 1994, despite the situation where **'group selection has been regarded as an anathema by nearly all evolutionary biologists'** (Richard Lewontin, 'Survival of the Nicest?', *The New York Review of Books*, 22 Oct. 1998), the biologist David Sloan (D.S.) Wilson desperately tried to **'re-introduce group selection…as an antidote to the rampant individualism we see in the human behavioral sciences'** (David Sloan Wilson & Elliot Sober, 'Re-Introducing Group Selection to the Human Behavioral Sciences', *Behavioral and Brain Sciences*, 1994, Vol.17, No.4).

[91] The right-wing then countered again, with biologist E.O. Wilson appropriating D.S. Wilson's Multilevel Selection theory that argued that natural selection operated at the group level as well as the individual level, thus accommodating an acknowledgement that we do have some selfless instincts, but still re-asserting the right-wing emphasis on selfishness. What left-wing biologists *then* did was try to bolster 'group selection' by adding the old 'by-products/many cranes/matrix of influences' illusion that Gould and Lewontin had first used. In effect, they threw everything into the pot: group selection and a multitude of vague 'influences' to emphasise that selflessness is supposedly an important part of our natural, genetic make-up! An example of this is provided in the 2011 book *Origins of Altruism and Cooperation* with editors Robert Sussman and Robert Cloninger writing in the Foreword that **'Research in a great diversity of scientific disciplines is revealing that there are many biological and behavioral mechanisms that humans and nonhuman primates use to reinforce pro-social or cooperative behavior. For example, there are specific neurobiological and hormonal mechanisms that support social behavior. There are also psychological, psychiatric, and cultural mechanisms'** (p.viii of 439). Yes, it was being alleged that a matrix of **'many biological and behavioral mechanisms'** created **'pro-social or cooperative behavior'**, but the question remains, how exactly did it do it? The illusion is that the origin of our moral instincts has been explained when it hasn't—but, again, in the desperation to assert the left-wing's selflessness-emphasising theory and counter the right-wing's selfishness-emphasising doctrine such extreme illusion was deemed necessary!

[92] Again, the reality under natural selection is, 'By all means you can help me reproduce my genes but I'm not about to help you reproduce yours'; it was only the extended nurturing of our infants that could overcome genetic selfishness and develop unconditionally selfless behaviour (see F. Essay 21).

[93] Basically, the biologically incorrect 'group selection' concept has been repeatedly employed by left-wing thinkers because even though they recognise it is a flawed idea, they haven't been able

to confront and acknowledge how nurturing was able to create our moral instincts, and as a result have had to keep reverting to, and attempting to bolster, the flawed concept of 'group selection' to assert that selflessness is a natural part of our make-up. [94] This stalled situation that so characterises biology now, where the Right use the false, savage, survival of the fittest instincts excuse to justify selfishness, and the Left use the false group selection plus a matrix of influences to try to say we are also naturally selfless, was perfectly captured in a 2010 documentary about these conflicting biological ideologies. Titled *Secrets of the Tribe* (about the Yanomamö Indians of the Amazon), the documentary ended up playing George and Ira Gershwin's well-known song *Let's Call the Whole Thing Off*, which features the lyrics, 'Things have come to a pretty pass…It looks as if we two will never be one…You like potato and I like potahto…Let's call the whole thing off.'

[95] I should mention that since *FREEDOM* was published in 2016, an elaboration on the fraudulent 'cooperation is more advantageous than competition and can therefore be selected for', 'group selection' theory has been developed. This elaboration argues that supposed selection against aggression within groups resulted in supposed increased friendliness within the group (a process that has been termed 'Self-domestication'), and brought with it an increased hostility towards 'outsiders'. For example, the explanation for why we hate others that was given in Steven Spielberg's 2019 documentary *Why We Hate* was based on this concept, which the anthropologist Brian Hare articulated in his 2020 book *Survival of the Friendliest: Understanding Our Origins and Rediscovering Our Common Humanity*: 'As humans became friendlier, we were able to make the shift from living in small bands of ten to fifteen individuals like the Neanderthals to living in larger groups of a hundred or more…But our friendliness has a dark side. When we feel that the group we love is threatened by a different social group…we are capable of… dehumaniz[ing] them…Incapable of empathizing with threatening outsiders, we can't see them as fellow humans and become capable of the worst forms of cruelty. We are both the most tolerant and the most merciless species on the

planet' (pp.XXVI-XXVII of 304). So this is the same fraudulent group selection type argument where we have some altruistic traits where we
are 'tolerant' and concerned for others within our group, along with
darker traits where we are selfishly, even 'merciless[ly]', concerned
for the reproduction of our own genes.

[96] Again, as I have emphasised and explained in *TI*, our selfless moral instincts were acquired through nurturing, not from
the biologically impossible 'group selection' mechanism. I also
emphasised in *TI* (and evidenced through references to the bonobos
and the fossil record) that our ape ancestors lived in a *completely*
cooperative and loving state, so our instinctive heritage is *not*
composed of some selfish and some selfless instincts but of entirely
unconditionally selfless instincts. Further, as I explain in FAQ 7.3
(which elaborates on what is explained in chapter 6:9 of *FREEDOM*),
developing upon the thinking of the philosopher Jean Jacques
Rousseau, the Left assert that the selfless side in us was eventually
overwhelmed by the supposed selfish side in us when, following
the advent of agriculture, settlements developed which allowed for
the accumulation of possessions; which the desire and competition
for supposedly then led to greed and warfare and domination by
the more powerful.

[97] So again, what I summarised in *TI* about what Sir Bob Geldof
said about the prevailing belief before the ascendency of Critical
Theory, that savage, must-reproduce-our-genes instincts dominate
our lives and so we have no choice but to dogmatically impose
selfless cooperation on society, is true. The only difference is that
the Right claim we have entirely savage, selfish, competitive and
aggressive, 'survival of the fittest', individualistic instincts, while
the Left claim that we do have some supposed group-selection-
derived, selfless, concern-for-others, social, communal instincts
influencing us along with these savage, selfish, competitive and
aggressive, 'survival of the fittest', individualistic instincts, with
the latter coming to dominate following the advent of agriculture,
sedentary living and the accumulation of possessions. Again, the
human-condition-confronting-not-avoiding, true explanation for

how our original completely cooperative and loving instinctive self or soul developed was through nurturing, and how it became corrupted was due to our conscious mind becoming competitive and aggressive when our instincts ignorantly criticised its necessary search for understanding.

[98] I should mention that because right-wing thinkers were also unable to confront the issue of the human condition and think truthfully about it and by so doing find the true, instinct vs intellect, reconciling, redeeming and rehabilitating explanation for why we corrupted our original instinctive self, they had no choice other than to use the false 'savage instincts' excuse to justify their defence of individualistic freedom. The classic example of this was Social Darwinism that misrepresented Charles Darwin's idea of natural selection as being a selfish, 'survival of the fittest' process. Darwin originally rightfully left it undecided as to whether those individuals that selfishly made sure they reproduced more could be viewed as winners, as being 'fitter', it was only later that others persuaded him to substitute the term 'natural selection' with the term 'survival of the fittest'. To very briefly explain why the 'survival of the fittest' concept is a misrepresentation of Darwin's idea of natural selection, and why Darwin was wrong to allow himself to be persuaded to use it: in F. Essay 23 and in chapter 4:2 of *FREEDOM*, I explain (as I mentioned earlier) that the meaning of existence is to develop ever larger and more stable wholes of matter, and that unconditionally selfless self-sacrifice for the good of the whole is the very theme of this Negative-Entropy-driven, integrative process because it maintains wholes. The fact that the gene-based natural selection process cannot normally develop unconditional selflessness between sexually reproducing individuals—because selfless traits don't tend to reproduce—is simply a limitation of the gene-based learning system; it does not mean that selfishness is the characteristic of existence. Integrative selflessness is the real characteristic of existence, the theme of life, which is why Darwin was wrong to allow the use of the selfishness-emphasising term 'survival of the fittest'. (For a more in-depth explanation of this and all the other distorted biological thinking that has been going

on as a result of biologists being unable to confront the issue of the human condition, see chapters 2, 4 and 6 of *FREEDOM*.)

[99] It is important to appreciate that since cooperative idealism is easy to justify as being a good and worthy attitude, the left-wing have had it easy saying that it's morally right, 'politically correct', to impose cooperative and loving behaviour. The Right, on the other hand, have had a very difficult job justifying the selfish, competitive and aggressive state of the angry, egocentric and alienated human condition. Yes, while the Left have found it harder to mount a biological argument that justified selfless behaviour, they have had an easier time arguing that selfless behaviour is good. And while the Right have had an easier job contriving a biological argument that justifies selfish behaviour, they have had an infinitely harder task trying to argue that selfish behaviour is justified. <u>Of course, the human-race-saving psychological relief that the instinct vs intellect Adam Stork analogy brings is that it finally truthfully explains the real reason for why selfish competition and aggression became an unavoidable, indeed necessary, part of human life—which does mean that we are now finally able to see that while both the Left and Right were using false biology, the ideology of the Left was wrong while the ideology of the Right was correct.</u>

[100] So when a journalist wrote that **'the great twin political problems of the age are the brutality of the right, and the dishonesty of the left'** (Geoffrey Wheatcroft, 'The year of sexual correctness and double standards', *The Australian Financial Review*, 29 Jan. 1999), we can now understand that he was referring to the right-wing's emphasis on the need to continue the upsetting, anger, egocentricity and alienation-producing, *brutal* search for knowledge free from the oppression of the *dishonesty* and delusion of the left-wing that maintained that we were being good by dogmatically imposing cooperative, selfless and loving ideal, 'politically correct' behaviour. Yes, we can finally clearly explain to the teenage climate change activist Greta Thunberg why it is actually her famous admonition of the right-wing that was **'all wrong'**, and **'not mature enough'**, and **'failing us'** (United Nations Climate Summit, 14 Dec. 2018).

[101] The following quote illustrates the complete ignorance of the Left of this need for individualism that the Right has championed. On the website *In Defence of Marxism*, the writer Ben Curry wrote in 2021 about Richard Lewontin, the Marxist biologist whose theories are included above, that **'In the 1970s, prominent individuals in the field of biology were once more taking up the reductionist philosophy that it's "all in our genes"…Lewontin understood that it was no accident that time and again these reactionary ideas penetrate the sciences. These ideas rest on a certain philosophical outlook, which derives from the outlook and interests of the ruling class. As Lewontin and [Richard] Levins explained in [their 1985 book]** ***The Dialectical Biologist,*** **their whole life they had fought "the mechanistic, reductionist, and positivist ideology that dominated our academic education and that pervades our intellectual environment."…The reductionist view sees the whole as nothing more than the sum of its parts. If we see war, greed and oppression in society, it is only because we are individually war-like, greedy and oppressive. In turn, we are each only an expression of our genes, which have evolved to make us this way because these traits give us a better chance of survival. These 'theories' suggest that our physiological, psychological and social traits are programmed into our genes, with a straight line of cause and effect between the latter and the former'** ('Richard Lewontin: the dialectical biologist (1929-2021)', 12 Jul. 2021).

[102] So although the ideology of the Right was correct, unable to confront the truth of our corrupted human condition and thus find the true instinct vs intellect explanation for why we have had to be competitive, selfish and aggressive, they were left having to contrive a way to justify our seemingly **'reactionary'**, non-ideal, competitive, selfish and aggressive behaviour, which they did by using the biologically dishonest Social Darwinist, 'survival of the fittest', 'savage instincts' excuse that said we humans are innately, naturally competitive, selfish and aggressive. Which meant the colossal philosophical heroes of the Right who bravely defended our competitive and aggressive state like the author Ayn Rand, and more recently the psychologist Jordan Peterson, were doomed to fail because their arguments were based on false biology. Rand

argued that we have to use our mind to effectively manage a selfish nature, and Peterson argues we have a competitive, dominance-hierarchy-based instinctive heritage that we have to acknowledge and manage rather than pretend doesn't exist, but immensely heroic as these two people have been, we can now understand that their biological thinking *was* fundamentally wrong. As an example of Peterson's use of false biology, he recently said about racism that it is **'a terribly deep human problem. In the 1970s Jane Goodall found that chimps go on raiding parties, and that was a major league discovery. It shows that the proclivity** [inclination] **to demonize the outgroup was at least 6 million years old. It's deep and it's in us, and we all have to contend with it'** ('Are All White People Racist?', Jordan B. Peterson YouTube channel, 6 Apr. 2022). As will be described later (see par. 124), our inclination to be intolerant of others is due to the psychological insecurity caused by our corrupted condition and has nothing to do with competing with outsiders to reproduce our genes. And it's peaceful bonobos not chimpanzees that are the model for our ancestors.

[103] Thank goodness we now have the true reconciling, redeeming and healing biological understanding of the human condition—which, incidentally, means the whole business of left and right wing politics is obsoleted. As summarised in paragraph 1136 of *FREEDOM* and mentioned in paragraph 118 of *TI*, the final irony of the saga of humanity's great journey from ignorance to enlightenment is that the ideal world that the left-wing was dogmatically demanding is actually brought about by the right-wing winning its reality-defending, freedom-from-idealism, corrupting-search-for-knowledge battle against the freedom-oppressing pseudo idealistic dogma of the left-wing. Yes, with the freedom-from-dogma right-wing's search for understanding of the human condition completed, the justification for the egocentric power-fame-fortune-and-glory-seeking way of life espoused by the right-wing ends, replaced by the ideal-behaviour-obeying attitude that the left-wing wanted. In this sense, when the right-wing wins we all become left-wing; through the

success of the philosophy of the right-wing, we all take up support of the ideal values sought by the philosophy of the left-wing—but, *most significantly*, this time we are not abandoning an ongoing battle, we are leaving it won. (For further reading on the fabulous obsolescence of politics, see F. Essay 34.)

- - - - - - - - - - - - - - - - - -

[104] Before going on to explain how the dishonest biological theories of both the Left and Right led to the rise of Critical Theory, I want to include another example of the Left's civilisation-destroying denial of the need for individualism.

[105] Following the federal election here in Australia in May 2022, which saw the conservative, right-wing 'Liberal' party thrown out of office, an absolutely appalling article by Richard Flanagan denouncing right-wing politics was published on 26 May 2022 in *The Sydney Morning Herald* (see www.wtmsources.com/299), which is one of our leading newspapers.

[106] With the outrageously deluded title **'Howard era ended with Morrison's downfall on the night Australia escaped its heart of darkness'**, the article shows how, when we haven't been able to admit humans once lived in a cooperative and loving innocent state (rather than as competitive and aggressive, must-reproduce-your-genes 'savages' as we've been taught) which we then corrupted and the human journey since then has been to search for understanding of *why* we corrupted ourselves, and instead dishonestly maintained we have unchangeable savage instincts that we have no choice but to dogmatically impose cooperative and loving ideal/politically correct behaviour on, it can be argued there's no justification for the right-wing—that John Howard (Australia's legendary former right-wing Prime Minister) and company are just brutal monsters from hell!!!

[107] What incredible delusion, dangerous madness in the extreme; again, it's dangerous because the left-wing's dogmatic imposition of ideal behaviour stifled the all-important corrupting search for know-

ledge, ultimately for self-knowledge, the rehabilitating understanding of ourselves that was needed to bring about a genuinely loving, peaceful world—which is unlike the right-wing that allowed the search for that all-important understanding to continue by tolerating a degree of competitive and selfish materialistic individualism.

[108] Yes, now that we can at last explain that we humans have been involved in an immensely upsetting but critically important heroic project to find understanding of ourselves, we can finally explain from first principle biology what is fundamentally wrong with the culture of the Left and fundamentally correct about the culture of the Right.

[109] Until now the only excuse the Right have had to justify their tolerance of a degree of competitive and selfish behaviour was to say, like Ayn Rand and Jordan Peterson have done, that since we supposedly have savage instincts everyone needs the motivation of winning some power, fame, fortune and glory if they are to participate in the supposed dominance hierarchy, law-of-the-jungle, survival-of-the-fittest, competitive individualistic world that we supposedly live in—that socialism doesn't work because it unrealistically kills people's incentive to be successful in, and thus actively participate in, a supposed competitive world.

[110] What an immense improvement it is to finally be able to explain the real reason for our competitive nature of the upsetting battle to find understanding of ourselves—which has finally been won, thus bringing about a reconciled and rehabilitated, truly peaceful world for humans. It's the Left that has been leading us into the **'heart of darkness'**, NOT the Right!

[111] The moral high ground, feel-good relief the article's author is deriving from slamming the Right as evil monsters is palpable. There is no interest in the possible merits of the Right, no interest in the human condition, no interest in the fundamental question of whether the divisive behaviour of the human race might be good and not bad after all.

10. The development of extremely dangerous Critical Theory

[112] So now to explain how the dishonest biological theories of both the Left and the Right led to the rise of the fraudulent, patently dishonest, extremely dangerous contrivance that is Critical Theory.

[113] Firstly, in terms of the progression of ever-increasing levels of upset that is being charted in this book, I described how the human race had arrived at the point where upset had become so extreme that it was decided we just had to leave the horrifically aggressive and selfish world we are living in and by any false means possible create a selfless and loving, human-condition-free world that the human race had always dreamed of achieving. As I summarised, the focus of politically correct post modernism was still mostly on relieving yourself of the guilt of your corrupted condition, so what was needed was a *program* that contrived a way that would allow the focus to be on the progression of the whole human race towards a more ideal state where all humans lived cooperatively instead of competitively.

[114] The question then was, what possible means could be invented to achieve that now desperately needed, dreamed-of transition? As just described, right-wing thinkers were arguing that it was only natural for humans to be competitive, selfish and aggressive because we supposedly have savage, 'must-reproduce-our-genes' instincts, and, to counter that individualistic, selfish view, left-wing thinkers argued that we are also naturally selfless and cooperative using the biologically impossible 'group selection' theory and, when that failed, adding the idea of a matrix of vague mechanisms to achieve it. In the end, these polarised, equally biologically flawed views proved unreconcilable, and biological thinking became stalled—to the point where the documentary suggested we should **'call the whole thing off'**.

[115] Unable to confront the human condition and find the reconciling and psychologically rehabilitating, true instinct vs intellect explan-

ation of our corrupted condition, and the true nurturing explanation for our selfless moral instincts, what left-wing thinkers did was revert to the ideology of Karl Marx who had, as previously mentioned, decided that **'The philosophers have only interpreted the world in various ways** [those various ways were the 'nature is red in tooth and claw' justification for selfishness *versus* spurious biological attempts to argue for selfless behaviour]; **the point is** [not try to understand our nature but] **to change it** [just make it cooperative/social/communal]'; as Marx also wrote, **'the alteration of men on a mass scale is necessary, an alteration which can only take place in a practical movement, a revolution'** (*The German Ideology*, 1845-1846). Basically, Marx had already recognised the problem of the 'red in tooth and claw', 'selfishness is the law of the jungle' world we live in, and because that is seemingly all science will tell us, we have to in effect bypass science and just dogmatically impose the cooperative, ideal world that is needed. What Marx realised is that the way to avoid this 'we are naturally savage, competitive and aggressive' idea was to limit any discussion of our selfish and selfless human nature to only acknowledging that we do all have some basic needs which, as the political theorist Norman Geras summarised in his 1983 book *Marx and Human Nature*, is **'for other human beings** [selflessness]**, for sexual relations, for food, water, clothing, shelter, rest and, more generally, for circumstances that are conducive to health** [selfishness]' (p.72 of 126). In fact, Marx took this 'all we have is some basic needs', ridiculously false idea to the extreme by maintaining that we are essentially born a 'blank slate', that our mind has no innate traits and can be inscribed at will. The fraudulent beauty of this idea that our **'nature'** was a **'self-creation'** (Karl Marx, *Economic and Philosophic Manuscripts of 1844*) was that Marx was then able to argue that it was social pressures that created our behaviour, and so all we need is a **'practical movement'** to **'change'** society in order to create a more loving world.

[116] So yes, when biology stalled in a polarised state of disagreement, left-wing thinkers went back to Marx's thinking and repurposed it

to achieve their mechanism for transforming the world into a more equitable and cooperative state—fake the arrival of a human-condition-liberated new world. Again, this left-wing thinking is **CRITICAL THEORY**. As the *Encyclopaedia Britannica* entry on Critical Theory says, '**Believing that science, like other forms of knowledge, has been used as an instrument of oppression, they caution against a blind faith in scientific progress, arguing that scientific knowledge must not be pursued as an end in itself without reference to the goal of human emancipation. Since the 1970s, critical theory has been immensely influential in the study of history, law, literature, and the social sciences**' (https://www.britannica.com/topic/critical-theory; accessed 19 Jul. 2021). And the *Stanford Encyclopedia of Philosophy* explains that '**a "critical" theory may be distinguished from a "traditional" theory according to a specific practical purpose: a theory is critical to the extent that it seeks human "emancipation from slavery", acts as a "liberating . . . influence", and works "to create a world which satisfies the needs and powers of" human beings...a critical theory provides the descriptive and normative basis for social inquiry aimed at decreasing domination and increasing freedom in all their forms**' (https://plato.stanford.edu/entries/critical-theory/; accessed 19 Jul. 2021).

[117]When Critical Theorists argue that '**Believing that science, like other forms of knowledge, has been used as an instrument of oppression**' and that '**they caution against a blind faith in scientific progress**', they are following Marx's view that '**The philosophers have only interpreted the world in various ways** [namely the 'nature is red in tooth and claw' justification for selfishness versus spurious biological attempts to argue for selfless behaviour]; **the point is** [not to try to understand our nature but] **to change it** [just create an equitable, cooperative/social/communal new world].' So they argue '**scientific knowledge must not be pursued as an end in itself without reference to the goal of human emancipation**'. They claim we have to bypass '**traditional**' biological thinking '**to create a world which satisfies the needs and powers of human beings**'. And since they blame our divisive behaviour on the '**oppressi**[**ve**]' **domination**' and '**slavery**' of social structures, we must '**free**' ourselves from those social constructs to achieve '**human emancipation**'.

[118] In short, Critical Theorists are Marxist in that they blame our corrupted condition on social constructs, in particular, capitalism. As Marx wrote, '**Under private property** [capitalism]**...Each tries to establish over the other an alien power, so as thereby to find satisfaction of his own selfish need. The increase in the quantity of objects is therefore accompanied by an extension of the realm of the alien powers to which man is subjected, and every new product represents a new potentiality of mutual swindling and mutual plundering**' (*Human Requirements and Division of Labour*, 1844). Marx is saying capitalism, the accumulation of possessions and wealth, caused us to become selfish, or at least perverted the fulfilment of our basic needs so that selfishness spiralled out of all control.

[119] Of course we can now understand that the truth is it was our upsetting battle with our instincts that unavoidably caused us to become selfish, not capitalism. Capitalism actually supplied us with the self-distracting materialism we needed while we heroically searched for self-understanding (see par. 111 of *TI*, and pars 289, 1092, 1119 & 1224 of *FREEDOM*). Our whole journey for the last 2 million years has been to find understanding of ourselves, find our meaning, find our identity, find understanding of why we are the way we are, namely competitive, aggressive and selfish when the ideals of life are so obviously to be cooperative, loving and selfless—basically, find the explanation of our corrupted, 'fallen' human condition. We needed to find that real validation of ourselves if we were to truly bring an end to the unavoidably upsetting angry, egocentric and alienated effects of searching for that understanding. As the philosopher Gerald Cohen said, '**Marxist philosophical anthropology**['s]**... conception of human nature and human good overlooks the need for self-identity...**[it] **underestimated the importance of phenomena, such as religion and nationalism, which satisfy the need for self-identity**' (*Reconsidering Historical Materialism*, 1978, p.163), and that '**there is a human need to which Marxist observation is commonly blind, one different from and as deep as the need to cultivate one's talents. It is the need to be able to say not what I can do, but who I am**' (ibid. p.348).

[120] So the next question is, having bypassed **'traditional' 'science',** how exactly did Critical Theorists contrive a way to achieve **'human emancipation'** and **'create a world which satisfies the needs and powers of human beings'**?

[121] What Critical Theorists did is not only call on the thinking of Marx, they even conscripted Sigmund Freud's analysis of our psychological state, which was very clever because they could then be seen to be addressing everyone's intuitive awareness that we suffer from psychosis! The entry on Critical Theory in the *Encyclopaedia Britannica* describes this sophisticated approach: **'Marxist-inspired movement in social and political philosophy originally associated with the work of the Frankfurt School** [in Germany]. **Drawing particularly on the thought of Karl Marx and Sigmund Freud, critical theorists maintain that a primary goal of philosophy is to understand and to help overcome the social structures through which people are dominated and oppressed.'**

[122] Marx had argued that workers tolerate the apparent injustice of the 'dominant ideology' of the status quo of capitalism, which encompasses the religious, political, economic and cultural aspects of an oppressive capitalistic society, because their mind is so dominated by it they can't see that they are being oppressed by it. The Frankfurt School, a movement founded in the 1920s by a group of left-wing social scientists, brought together Marx's idea of controlling 'dominant ideology' with the theories of Sigmund Freud, who had discovered that humans have a 'personal unconscious', a repository of repressed thoughts and emotions that are too painful to acknowledge consciously, including supposedly unacceptable primitive instinctive urges such as sex and aggression—all of which supposedly subverts our conscious thinking. The Frankfurt School argued that the reason that oppressed people around the world had not risen up in a great 'Marxist revolution' against supposedly unjustly oppressive structures like capitalism was because this 'false consciousness' created by both society and their own repressed unconscious prevented them from seeing the true oppressive nature of the society they laboured

under. Therefore, in order for the masses to rise up, the Frankfurt School worked on a system that would teach people to overcome 'false consciousness' by thinking critically about everything they had previously taken for granted. The result being Critical Theory, which sought to identify the hidden biases within standard modes of thinking, such as classism, racism and sexism, that supposedly allow for continued abuses of power. Once people could see through their 'false consciousness' and identify their 'oppressors', then the 'revolution', which was the forcible and dogmatic imposition of cooperative ideals, would supposedly naturally occur. As Max Horkheimer, one of the school's founders, wrote, the ultimate goal of Critical Theory was **'to liberate human beings from the circumstances that enslave them'** (*Critical Theory: Selected Essays*, 1982, p. 244).

[123] So the Frankfurt School combined Marx's idea of a controlling 'dominant ideology' of capitalism, which encompasses the religious, political, economic and cultural aspects of capitalistic society, with Freud's discovery of a 'personal unconscious' which supposedly subverts our conscious mind, to supposedly explain why we have a 'false consciousness' that contains hidden biases like classist, racist and sexist thoughts—which they then attempt to dogmatically eradicate in ourselves and in society. As I mentioned, capitalism actually supplied the self-distracting materialism we needed in order to carry on our corrupting search for knowledge, ultimately for self-understanding. In the case of religion, as has already been explained, when we humans became overly upset from participating in humanity's heroic battle to find knowledge, we needed to be able to defer to religious faiths because they offered relief from our own corrupted condition.

[124] With regard to Freud's theory of the 'personal unconscious', Freud did recognise that we have a repressed unconscious or psychosis that affects our ability to think truthfully, which we can now understand is all the denial we have had to practice of our 2-million-year corrupted condition while we couldn't truthfully explain it. We are enormously, immensely, astronomically ashamed and thus

insecure about our corrupted condition and as a result practice blocking out from our mind the truth of our original all-loving and all-sensitive instinctive self or soul; the result of which is, as Freud bravely recognised, the 'repressed unconscious' part of ourselves. And the extent of this denial or block-out or alienation or soul-repression or psychosis obviously will vary according to how exposed we have each been in our life to all the upset in the world. It follows that (as I describe in chapter 8:16E of *FREEDOM*, and in F. Essay 28) there are differences in alienation between individuals, races, genders, ages, generations, countries, civilisations and cultures. And as a result of these differences we do all naturally suffer from insecurity about our particular state of alienation compared to other people's states of alienation, and so we are all naturally variously prejudiced—the more innocent are prejudiced against the more alienated, who to them seem to be 'bad', and the more alienated have retaliatory prejudice towards the less alienated for their either direct or implied condemnation of them; and so we are all prone to being classist, racist, sexist, ageist, and so on. But again, the only way to end all that insecurity within us was to find the healing understanding of our corrupted human condition. The dogmatic imposition of ideality on our upset reality just added more denial/alienation/unconscious bias—and worse still, it kills the freedom we need to be able to continue the upsetting search for knowledge in order to find the psychologically redeeming and relieving understanding of ourselves.

[125] (In my book *How Laurens van der Post Saved The World* I describe and explain at some length the journey of acknowledgement of our corrupted condition that Freud bravely initiated with his recognition of a 'repressed unconscious' part of ourselves, and how Freud's associate, Carl Jung, then bravely added to Freud's recognition of a 'repressed unconscious' within us by recognising the foundation insecurity within that 'repressed unconscious' of a repressed awareness of a collective, shared-by-all instinct within us that Jung termed our 'collective unconscious'. Sir Laurens van der

Post and myself added the critical clarification that our 'collective unconscious' is actually the repressed instinctive memory within us all of a time when our ape ancestors lived in a completely co-operative, selfless and loving state prior to becoming conscious and developing our present corrupted condition—a memory we repressed because we couldn't explain why we had destroyed such a wonderful, all-loving and all-sensitive state.)

[126] Yes, what was so extremely dangerous about Critical Theory's overall practice of dogmatically imposing a cooperative, ideal world was that the dogmatic imposition of selfless, cooperative and loving ideal/politically correct behaviour oppresses and stifles the all-important freedom of expression needed to find knowledge, ultimately the all-important self-knowledge, the redeeming, rehabilitating and transforming biological understanding of our psychologically distressed human condition that *actually* brings about the cooperative, selfless and loving world that is so desperately needed.

[127] The great paradox of the human condition (which is what the Adam Stork analogy finally explains) is that we have had to be free from our dogmatically oppressive and dictatorial cooperation-and-selflessness-demanding moral instinctive self or soul if we were to carry on our soul-corrupting, anger-egocentricity-and-alienation-producing, competition-and-selfishness-occurring, heroic search for knowledge. Again, in the words from the song *The Impossible Dream* from the musical the *Man of La Mancha*, we had to be prepared to **'march into hell for a heavenly cause'** (lyrics by Joe Darion, 1965); we had to be prepared to lose ourselves if we were to find ourselves; we had to suffer becoming angry, egocentric and alienated until we found sufficient knowledge to explain ourselves and by so doing actually end and heal our corrupted condition.

[128] So, as I mentioned earlier, dogma is not the cure for the troubled state of the world, it is in fact the *poison* because it blocks the search for the rehabilitating understanding of ourselves that is needed to *actually* save the world. The truth is the Left is *not* **'progressive'** as it deludes itself it is, but regressive. It was actually

the right-wing who supported the upsetting anger-egocentricity-and-alienation-producing battle to find knowledge that held the moral high ground, not the pseudo idealistic left-wing. The real way to bring an end to our selfish and divisive condition where **'people are dominated and oppressed'** and **'create a world which satisfies the needs and powers of human beings'** and achieves **'the goal of human emancipation'** and **'freedom'** from our corrupted angry, egocentric and alienated human condition depended on continuing the upsetting search for knowledge until we found the true, psychologically redeeming and rehabilitating, instinct vs intellect biological explanation of the human condition—which, thank heavens, has finally been found, but is yet to be widely acknowledged. So again, Marx was wrong when he said, **'The philosophers have only interpreted the world in various ways; the point is** [not to understand the world but] **to change it** [just make it cooperative/social/communal]', because the whole **'point'** and responsibility of being a conscious being *is* to understand our world and our place in it—ultimately, to find the redeeming and psychologically healing understanding of our seemingly horribly flawed, 2-million-year upset, now totally mad and deranged, human condition. We humans needed answers for our upset, distressed mind, not dogma; we needed brain food not brain anaesthetic; we needed to be able to think our way to sanity, not become brain-dead robots. De-braining ourselves was never going to work. To avoid the fast approaching terminal levels of alienation/psychosis we simply had to find the psychologically healing understanding of ourselves.

[129] When the author Antoine de Saint-Exupéry wrote that **'We are living through deeply anxious days, and if we are to relieve our anxiety we must diagnose its cause…What is the meaning of man? To this question no answer is being offered, and I have the feeling that we are moving toward the darkest era our world has ever known'** (*A Sense of Life*, pub. 1965, pp.127, 219 of 231), he was recognising that the survival of our species depended on finding understanding of ourselves. So to effectively stop that search was the ultimate crime against the whole conscious-thinking

human race! The truth is the left-wing culture of dogmatically en-
forcing cooperative, selfless and loving behaviour has been leading
humanity not to freedom from the human condition as it deludes
itself, but straight off the cliff of hope and down into the chasm of
extinction; to death by dogma!

11. Critical Theory's dangerous derivatives of Critical Race Theory and Critical Gender Theory

[130] So that is the explanation of what's so very wrong and danger-ous about Critical Theory, which, as I will now describe, means its offshoots of Critical Race Theory (CRT) and Critical Gender Theory—and their manifestations of 'Identity Politics', 'Woke' ideology, 'Cancel Culture', and the 'Great Reset' of society—are also very wrong and very dangerous.

[131] The question that arises is how was Critical Theory—that so dangerously decided we have no choice other than to artificially impose cooperative and loving behaviour on science's supposed finding that we are naturally selfish creatures living in a naturally competitive and selfish world—actually going to be implemented? The answer to this question is that once it was decided we had to stop looking to science for guidance, we were free to assert whatever we liked, and that's exactly what happened.

[132] To impose a new world of love and kindness between ethnic groups or races, it was simply asserted that it was philosophically sound to claim that there is no difference between races; and further that any contention that there were differences was just a dishonest, manipulative, racist, artificially invented device used to oppress and ill-treat certain races. Essentially, there was no recognition at all that humans are involved in a grand project that has inevitably resulted in everyone and every related group of humans being differently upset by their inevitable different encounters with that upsetting grand project (see F. Essay 28). No human condition to find understanding of, just dogmatically **'change'** our behaviour, as Marx said. The product of this thinking was **CRITICAL RACE THEORY (CRT)** which is described as an **'intellectual movement and loosely organized framework of legal analysis based on the premise that race is not a natural, biologically grounded feature of physically distinct subgroups of human beings but a socially constructed (culturally invented) category that is used to oppress and exploit people of colour. Critical Race Theorists hold**

that the law and legal institutions in the United States are inherently racist insofar as they function to create and maintain social, economic, and political inequalities between whites and nonwhites, especially African Americans' (https://www.britannica.com/topic/critical-race-theory; accessed 19 Jul. 2021).

[133] And furthermore, while feminism tried to impose equality between the sexes by in effect denying that any real differences existed between them, Critical Theory sought to reinforce that idea by inventing a philosophy that supposedly firmly established that there weren't any meaningful differences. To do this it was asserted that while there are physical differences between the sexes there is no significant psychological differences between them; and it was asserted that any argument that there was, was simply a dishonest, manipulative, sexist device used to oppress and ill-treat one or other of the genders, but women particularly. Basically, any suggestion was denied that in the human journey a different role for men would have inevitably resulted from men being the group protectors and as a result being most responsible for taking on the immensely upsetting human-race-defending job of defying our condemning instincts in the all-important search for knowledge—and that a different role for women would have inevitably resulted from women being the bearers of offspring and not as responsible as men for the immensely upsetting human-race-defending job of defying our condemning instincts, with their focus having to be on the next generation, of giving birth to them, suckling them, sheltering them from upset as much as they could, and nurturing them with as much upset-free, unconditional love as they can manage in an extremely upset world (see F. Essay 27). Different roles for men and women, and the different psychological effects those roles have, and the different psychological strategies that men and women could adopt to cope with the potentially extreme agonies of their situations, were all denied. The product of this thinking was **CRITICAL GENDER THEORY**, where **'critical gender theorists attack the assumption that every human being naturally belongs to one of two discrete gender categories (masculine or feminine), which is determined by biologically-given sexual**

characteristics (male and female). Rather, according to critical gender theory, there is no necessary connection between biological sex and a person's gender presentation. The ways in which men and women present themselves as gendered individuals are social constructs, learned performance, and "a social accomplishment." Biology has no causal influence on gender...According to critical gender theory, while one *may* be born a sexed being, one is not born gendered. One must learn gender presentation' (Terry S. Kogan, 'Transsexuals and Critical Gender Theory', *Hastings Law Journal*, 1997).

[134] Since a completely sound philosophical foundation had supposedly been established for these critical theories, any deviation from them was supposedly illegitimate and thus to be admonished and shut down, cancelled or eliminated or removed, which is what is referred to as **CANCEL CULTURE**. Further, since these critical theories were supposedly based on solid philosophical foundations, if you didn't subscribe to them, you were in effect still 'asleep' living in an obsolete paradigm; you weren't **WOKE** and part of the great awakening to a cooperative and loving new world; you weren't part of the supposed **GREAT RESET** of humanity from living in a competitive and aggressive condition to **BUILDING BACK BETTER** to living in a cooperative and loving one that was free of that supposedly immensely divisive and wrong behaviour. And **IDENTITY POLITICS** developed where those suffering most from the great battle humanity has actually been waging to find understanding of the human condition could supposedly legitimately demand 'equity' of recognition, 'inclusion' in material prosperity and 'diversity' of representation. There was no recognition of the human race's great project to find the knowledge that would actually liberate us from our corrupted condition, and how involvement in that great project unavoidably created various states of upset and functionality—where, for example, those presently most functional under the duress of the human condition, and the gender most responsible for searching for knowledge, namely white males, had to and have succeeded in finding the liberating understanding of the human condition, as all the great thinkers mentioned in this book evidence. So it has

not been a case of unfair white privilege, especially unjust **WHITE MALE PRIVILEGE** as critical theorists are teaching, but rather a case of fulfilled white male responsibility. The 'pale, stale males' are actually the heroes, legends and saviours! (See F. Essay 28 and chapters 8:16E-F of *FREEDOM* for an explanation of the differences in alienation between races which—now that the redeeming explanation of the human condition has been found—is able to be safely and responsibly acknowledged.) Instead of such truthful analysis of our situation, there's just been Marxism's dogmatic and deluded demand that everyone behave cooperatively and lovingly—such that, supposedly, as is being said by its advocates, 'You'll own nothing and you'll be happy.' What an outrageously dishonest fraud! How could we possibly 'be happy' living without the psychologically redeeming, reconciling and healing understanding of our 2-million-year corrupted condition that our minds have so desperately needed and have so heroically sought! What deluded madness to claim that our world is being made better when in fact the human race is being taken headlong to a state of terminal denial/alienation/psychosis and extinction!

12. Terminal alienation and the extinction of the human race was upon us

[135] The essential truth that we can now understand is that the source problem was the inability to confront the issue of the human condition and by so doing find the true instinct vs intellect rehabilitating understanding of it. Failure trapped by that inability, in sheer desperation because of all the unbearable levels of upset, a no-healing-understanding-for-our-2-million-year-psychologically-upset mind, totally artificially and ineffectively transformed world where humans supposedly live cooperatively and lovingly instead of competitively and aggressively was invented and tyrannically imposed.

[136] The gloves were off now, the confidence of—and the sheer anger, aggression and fury underlying—the industry of denial and delusion was such that it was now prepared to go the whole hog and brazenly mimic the arrival of the human-condition-understood-and-reconciled true world that the human race has always dreamed of achieving, cruelling any chance of it actually arriving. The fact is, the postmodern, politically correct, critical theory culture, with its Critical Race and Gender Theories, Identity Politics, Wokeness, Cancel Culture and Great Reset agenda, represented the very height of dishonesty, the most sophisticated expression of denial and delusion to have developed on Earth. Terminal alienation was indeed upon us, an extinction of our species *was* just around the corner!

[137] It has been utterly unbearable for us humans having to live without the redeeming explanation of our corrupted, soul-destroyed, Garden-of-Eden-state-of-original-innocence-devastated, human condition while we searched for that relieving understanding, all the time becoming more and more psychologically upset as the search went on. The great danger was always that eventually we would become so *extremely* upset/soul-corrupted that the feelings of unworthiness, shame and guilt would become *so* great that our minds would become fanatically attached to whatever artificial forms of relief we had been able to find and develop, no matter how dishonest, deluded and mad those forms of relief were. After many stages of progression that have been described in this book, the human race has finally arrived at this extreme situation where artificial, do-'good'-to-feel-good pseudo idealism, now in the form of the politically correct, dogmatic lunacy of the left-wing, is fanatically seeking to impose itself on the world. And it's happening at the highest levels of society, in universities, schools, science, the judiciary, politics, medicine, the media, the church, royalty, and the corporate world, especially in the monolithic technology companies. These are all institutions that have traditionally been politically impartial. There is literally no sanity left, just psychological sickness and its desperate madness everywhere.

[138] A 2017 article by Matt Ridley in *The Times*, titled 'Blazing Saddles a dim memory as society enters a new dark age', describes how **'The enforcement of dogma is happening everywhere'**, **'We are witnessing the sabotage of the core principles of a free society'**, **'There's an almost religious quality to many of the protests'**. Ridley also refers to **'virtue signalling** [which is signalling how virtuous you are], **written with the sanctimonious purity of a Red Guard during China's Cultural Revolution'**, and to the present **'safe-space, trigger warning culture'**, saying that **'maybe the entire world is heading into a great endarkenment'**. He ends by saying **'the intolerance of dissent in Western universities and the puritanical hectoring of social media give grounds for concern that the flowering of freedom in the past several centuries may come under threat'** (26 Sep. 2017). Yes, these cautionary words from the philosopher John Stuart Mill have never been more applicable: **'We have now recognised the necessity to the mental well-being of mankind (on which all their other well-being depends) of freedom of opinion, and freedom of the expression of opinion'**, for **'the price paid for intellectual pacification, is the sacrifice of the entire moral courage of the human mind'** (*On Liberty*, 1859).

The celebrated Australian cartoonist Bill Leak (1956–2017) depicted the anger and intolerance at the heart of the pseudo idealism of political correctness.

[139] As I summarised in *TI*, the culture of the Left makes people superficially feel good but it is extremely dangerously dishonest. Being concerned for others and the world is *very* important, but doing that to make yourself feel good is a dangerously selfish sickness, indeed it's the most seductive and destructive of all drug addictions—and it's been taking over the world. Dogma is *not* the cure, it's the poison.

[140] Yes, it is a case of getting artificial feel-good relief from the unbearable guilt of the human condition at all costs, and it *is* at 'all costs' because it meant the human race was abandoning the all-important heroic search for knowledge and any chance of finding the understanding of the human condition that would *actually* end the agony of the human condition, which thankfully has arrived in the nick of time to save the world from this horrific death by dogma—as long as we can get this critical breakthrough understanding of the human condition supported before we are outrun by all the habituated denial and attachments to all the useless ways we are trying to save ourselves. In fact, not only did left-wing pseudo idealism hinder the finding of understanding we so desperately needed, now that it has been found, the Left's prohibition on any profound thinking means that it will try to ensure it doesn't see the light of day! So please support, help and join the World Transformation Movement that, often against ferocious resistance, is promoting the instinct vs intellect explanation of the human condition—the only thing that can end all the suffering and save humankind from extinction.

[141] So, we can now fully understand from a basis of truthful, first principle biology how dangerous pseudo idealism has been, and appreciate the philosopher Friedrich Nietzsche's warnings that **'There have always been many sickly people among those who invent fables and long for God** [ideality]**: they have a raging hate for the enlightened man and for that youngest of virtues which is called honesty...Purer and more honest of speech is the healthy body, perfect and square-built: and it speaks**

of the meaning of the earth [which is to fight for knowledge, ultimately self-knowledge, understanding of the human condition]...**You are not yet free, you still** *search* **for freedom. Your search has fatigued you...But, by my love and hope I entreat you: do not reject the hero in your soul! Keep holy your highest hope!...**War [against oppression, especially from dogma] **and courage have done more great things than charity. Not your pity but your bravery has saved the unfortunate up to now'** (*Thus Spoke Zarathustra: A Book for Everyone and No One*, 1892; tr. R.J. Hollingdale, 1961, pp.61-75 of 343), and **'There comes a time in a culture's history when it becomes so pathologically soft that it takes the side of its worst enemy** [dogma]**...and calls it "progress"'** (common tr. of *Beyond Good And Evil*, 1886, sec. 201).

[142] The science historian Jacob Bronowski gave a similar warning about not becoming **'soft'** in his concluding statement to his 1973 television series and book of the same name, *The Ascent of Man*: **'I am infinitely saddened to find myself suddenly surrounded in the west by a sense of terrible loss of nerve, a retreat from knowledge into – into what? Into... falsely profound questions about, Are we not really just animals at bottom; into extra-sensory perception and mystery. They do not lie along the line of what we are now able to know if we devote ourselves to it: an understanding of man himself. We are nature's unique experiment to make the rational intelligence prove itself sounder than the reflex** [instinct]**. Knowledge is our destiny. Self-knowledge, at last bringing together the experience of the arts** [that describe the human condition] **and the explanations of science** [that has to explain the human condition], **waits ahead of us'** (p.437 of 448).

[143] Basically, the virtuous, righteous, pseudo idealistic causes we humans have taken up after becoming **'fatigued'**, such as socialism, new ageism, feminism, environmentalism, political correctness, post-modernism, Marxist Critical Theory, identity politics, woke-ness, cancel culture and the Great Reset, and so on, all represented *false starts* to a human-condition-free world—because, once again, the *real start* to an anger, egocentricity and alienation-free world depended on continuing the upsetting search for knowledge until we found the reconciling understanding of the human condition. As Bronowski and Nietzsche respectively said, **'Knowledge is our destiny'**

and the anti-knowledge, anti-truth and anti-progress of dogma is **'its worst enemy'.**
[144] And, as Socrates famously pronounced, **'the only good is knowledge and the only evil is ignorance'** (Diogenes Laertius, *Lives of Eminent Philosophers*, c.225 AD), and **'the unexamined life is not worth living'** (Plato's dialogue *Apology*, c.380 BC; tr. B. Jowett, 1871, 38) — but in the end a preference for ignorance and the associated need to oppress any examination of our lives, oppress any freedom to think truthfully, question and pursue knowledge, threatened to become the dominant attitude throughout the world. George Orwell's bleak prediction in his famous 1949 book *Nineteen Eighty-Four* that **'If you want a picture of the future, imagine a boot stamping on a human face** [the human mind] **– for ever'** was about to be realised.

[145] It's always been prophesised that when understanding of the human condition arrives it will defeat a terrible false liberator of humanity, a false messiah, which we can now understand is pseudo idealism. Islam maintains that the 'messiah' (which I explain in par. 1278 of *FREEDOM* is science assisted in the end by denial-free thinking) will **'defeat** *Al-Masih ad-Dajjal*, **the false messiah'**, the **'evil deceiver'**

who will **'try to lure people'** and **'fool'** everyone (Wikipedia). In par. 1126 I describe how Christ also warned of **'the end of the age'** when **'many false prophets will appear and deceive many people…Wherever there is a carcass** [the extremely upset], **there the vultures** [false prophet merchants of delusion and denial] **will gather…So when you see the "abomination that causes desolation," spoken of through the prophet Daniel, standing where it does not belong** [pretending to be bringing about an enlightened 'woke' world]**…flee to the mountains…How dreadful it will be in those days…If those days had not been cut short** [by the arrival of the real liberator of understanding of the human condition]**, no-one would survive.'** You can't have a clearer or more authoritative warning of the danger of pseudo idealism than that—'authoritative' because Christ is one of the soundest denial-free, honest, effective thinkers in history, see F. Essay 39. (It is true that religion itself is a form of pseudo idealism, of artificially imposing ideality on our corrupted condition, but, as I have explained, it was by far the least dishonest and deluded of the forms of pseudo idealism that developed after it.)

[146] I might also include another strong warning of the danger of pseudo idealism from yet another great prophet, Sir Laurens van der Post; indeed, his full-page obituary in London's *The Times* was boldly headed **'A Prophet Out of Africa'** (20 Dec. 1996; see www.wtmsources.com/166): **'the so-called liberal socialist elements in modern society are profoundly decadent today because they are not honest with themselves…They give people an ideological and not a real idea of what life should be about, and this is immoral…They feel good by being highly moral about other people's lives, and this is immoral… They have parted company with reality in the name of idealism…there is this enormous trend which accompanies industrialized societies, which is to produce a kind of collective man who becomes indifferent to the individual values: real societies depend for their renewal and creation on individuals…There is, in fact, a very disturbing, pathological element—something totally non-rational—in the criticism of the** [then UK] **Prime Minister** [Margaret Thatcher]**. It amazes me how no one recognizes how shrill, hysterical and out of control a phenomenon it is…I think socialism, which has a nineteenth-century inspiration and was valid really only in a nineteenth-century context when the working classes had**

no vote, has long since been out of date and been like a rotting corpse whose smell in our midst has tainted the political atmosphere far too long' (*A Walk with a White Bushman*, 1986, pp.90-93 of 326).

13. The solution

[147] How precious is it then that understanding of the human condition has been found and this terrifying imminent prospect of the death-by-dogma end of the human race can be avoided, and the *real* transformation of the human race from a self-preoccupied, selfish, human-condition-stricken state to a concerned-for-others, selfless, human-condition-understood, psychologically rehabilitated state can finally occur?!

[148] So please support our World Transformation Movement that promotes this denial-free new science of the redeeming, reconciling and rehabilitating biological explanation of the human condition because all other projects are futile and only this project can save the world.

Graphic by J. Griffith, M. Rowell and G. Salter © 2009 Fedmex Pty Ltd

Addendum 1 – Humanity is coming home

[149]Civilised freedom of expression and individualism, sustained by the incentives of materialism and capitalism, are what maintained humanity's heroic search for knowledge, ultimately for self-knowledge, the psychologically rehabilitating understanding of the human condition, that now allows for humanity's joyous return home to soundness.

Understanding leads to the utopia of the rehabilitation of humans

[150]Whereas Marxism's pseudo idealistic, dogmatic imposition of politically correct, cooperative and selfless behaviour that is supposed to bring about a better world actually stifled and oppressed the freedom of expression and individualism needed to find the psychologically rehabilitating understanding of the human condition, and so only led to terminal alienation and the homeless and lawless end of civilisation.

Dogma leads to the dystopia of lawlessness, unbearable psychosis and homelessness

[151] Having in the last century taken up the position at the forefront of humanity's heroic search for knowledge—being, as its national anthem says, 'the land of the free and the home of the brave'—means the United States will also be suffering greatly from the psychological upset that that search produces, like countries that previously led that heroic search have already experienced in the 'peaked to decadent' cycle. And with mass communication being so powerful now, the whole world has been influenced by the United States' great burst of creativity and the accompanying escalation in psychosis, which means that the world would follow what is happening in the United States if understanding of the human condition hadn't been found. And what we see happening there, especially in states like California that are at the cutting edge of the United States' 'progress', is the adoption of the most advanced forms of 'makes-you-feel-good-but-is-actually-extremely-deluded-and-dangerous' pseudo idealism to artificially relieve psychosis. The land of the free

and brave is becoming the land of the psychologically exhausted where the jackboot of dogma is leading to a zombie world of soul-dead darkness and human extinction. The **'love-in there in the streets of San Francisco, gentle people with flowers in their hair'** sung about only a short while ago in 1967 by Scott McKenzie has become the defiled streets of psychotic people with torture in their brains. **'California dreamin'** (*The Mamas & the Papas*, 1965) is now California leavin'! San Francisco is being referred to as 'Sanfransicko', and being at the leading edge of where the rest of the world is heading, 'SICK' is what the whole human race is plunging towards! What a relief then that the rehabilitating, transforming understanding of the human condition has been found!

Zombie street parades (see video in F. Essay 1), which have become popular of late, are an exorcism, a letting out, of the truth that the human race is entering the endgame state of terminal psychosis.

[152] A former speaker of the US House of Representatives, Newt Gingrich, spoke the truth about what has really been going on when he said about the Left that '**these are people who need therapy. We try to deal with this as though it's a political problem. It's not. It's a mental health problem. These people are crazy**' (*Hannity*, Fox News, 20 Dec. 2021). The irony is that what the Left are doing *is* therapy, they are trying to relieve themselves of extreme human-condition-stricken agony by finding causes that make them feel good about themselves. The problem is it's dangerously dishonest *artificial* therapy, not the *real* 'therapy' that Gingrich is recognising they need — which, mercifully, is now available.

[153] The horrific levels of delusion and selfishness involved in the real-progress-towards-understanding-ending, civilisation-destroying culture of the PC, woke Left was captured by Elon Musk, the richest man in the world's famous comments that '**At its heart, wokeness is divisive and exclusionary, and hateful. It basically gives mean** [psychologically upset] **people a shield to be mean and cruel, armored in false virtue**' (2021), and the '**Woke mind virus is the biggest threat to civilization**' (2022). We can only hope that Musk ends up buying Twitter! But, as I've carefully described in this book, the desperate need for artificial relief from the agony of the human condition has become so great that such extremely dangerous delusion has absolutely reached its end point! The philosopher Antonio Gramsci got it right when he wrote that '**The crisis consists precisely in the fact that the old is dying and the new cannot be born; in this interregnum a great variety of morbid symptoms appears**' (*Prison Notebooks*, 1927-1937). The breakthrough of someone finally confronting the unconfrontable issue of the human condition and by so doing allowing to '**be born**' the understanding of our corrupted condition has come not a moment too soon!

[154] With regard to Musk's SpaceX enterprise, I would like to say to Elon that the real frontier is not outer space but inner space, the issue of the human condition. Outer space was a necessary distracting escape, but the way forward now that understanding of the human condition has been found is back to our soul's

world of soundness, real togetherness and real happiness. Also, I would like to make the point about Elon's brilliant technological innovations, and about the advances being made in Artificial Intelligence, which I know Elon is concerned about, that playing God with ever more sophisticated technology while we couldn't even confront God (Integrative Meaning) and were living in complete denial of how psychotic we are, was a potentially very dangerous pursuit.

[155] Finally, with regard to the finding of this human-race-saving understanding of the human condition occurring in Australia — due to its relative isolation from all the upset in the world, Australia really is the last bastion of the innocence needed to take up the role from the United States of continuing humanity's heroic search for knowledge, and (with innocence in Australia, and time in general, rapidly running out) hopefully taking that search to its conclusion, as it in fact has. In chapter 9:11 of *FREEDOM* I have written at some length about this critical role of Australia. Most extraordinarily, I include in paragraphs 1265-1272 the Nobel Laureate Albert Camus's 1940 essay *The Almond Trees* in which Camus all but says that first principle, reconciling biological explanation of the human condition will be found in Australia. He states that 'men have never ceased to grow in the knowledge of their destiny', which is to 'overcome our condition', 'find those first few principles' that will 'stitch up what has been torn apart'. He then prophetically writes how we must not give in 'to despair', that despite 'nations poisoned by the misery of this century…[who have] utterly surrendered to that evil which Nietzsche called the spirit of heaviness [psychosis/depression]', and 'the vastness of the undertaking', 'I turn towards those shining lands where so much strength is still untouched. I know them too well not to realize that they are the chosen lands where courage [to defy all the dishonest denial in the world] and contemplation [denial-free, honest thought] can live in harmony.' I know that 'through the virtue of its whiteness and sap, [its innocent soundness, it] stands up to all the winds from the sea [stands up to the immensely dishonest and deluded pseudo idealism in the world]. It is that which, in the winter for the world, will prepare the fruit [solve the human condition and bring humanity home to peace and sanity].'

Addendum 2 – Assault on the West explained

[156] I've decided it might be very helpful to add a second Addendum, this one explaining why the Judeo-Christian Western culture, like the right-wing in politics, has been one of the main targets for rejection by the pseudo idealistic, politically correct, 'woke', identity politics, cancel culture movement.

[157] The British author and political commentator Douglas Murray gave a good description of this rejection of the Western tradition in a summary he wrote of his 2022 book *The War on the West*: **'In a few short decades, the Western tradition has moved from being celebrated... to being something shameful...while the West is assaulted for everything it has done wrong, it now gets no credit for having got anything right. In fact, these things – including the development of individual rights, religious liberty, and pluralism – are held against it...The culture that gave the world lifesaving advances in science, medicine and a free market that has raised billions of people around the world out of poverty and offered the greatest flowering of thought anywhere in the world is interrogated through a lens of the deepest hostility...All aspects of the Western tradition now suffer the same attack. The Judeo-Christian tradition that formed a cornerstone of the Western tradition finds itself under particular assault and denigration. But so does the tradition of secularism and the Enlightenment, which produced a flourishing in politics, sciences and the arts...these inheritances are criticised as examples of Western arrogance, elitism, and undeserved superiority. As a result, everything connected with the Western tradition is being jettisoned. At education colleges in America, aspiring teachers have been given training seminars where they are taught that even the term "diversity of opinion" is "white supremacist bullshit"...this war on the West...now exists at the very top of the American government, where one of the first acts of the new administration was to issue an executive order calling for "equity" and the dismantling of what it called "systemic racism". We appear to be in the process of killing the goose that has laid some very golden eggs'** (*The Australian*, 30 May 2022).

[158] Yes, the Judeo-Christian Western tradition *has* 'laid some very golden eggs', BUT, the problem is, it seemingly hasn't laid THE 'golden egg' it was entrusted to find, namely the reconciling, redeeming and human-race-liberating understanding of the human condition. I say 'seemingly hasn't' because understanding of the human condition *has* actually been found and is presented in my book *FREEDOM*, but that breakthrough is yet to be widely known.

[159] The explanation of the significance of this seeming failure to solve the human condition by the West begins with recognition of the fundamental, extremely difficult situation the human race has been in. This situation is that while the good reason for the corrupted state of the human condition wasn't able to be explained it was too unbearably condemning and depressing to admit the truth of the existence of that corrupted state—and not being able to admit it, it obviously wasn't possible to acknowledge that the fundamental objective of the human race was to find the redeeming explanation for why we corrupted our species' original innocent state. We couldn't acknowledge our corrupted condition and so couldn't admit that our fundamental task was to find the explanation for why we corrupted it. Nevertheless—and this is very important—while we've had to live in Plato's metaphorical dark cave of denial of our corrupted condition and task of solving it, all humans *have* carried a deep intuitive awareness of the truth that we do suffer from the corrupted state of the human condition and that our fundamental task has been to find the redeeming explanation for why we became corrupted. All truly great literature, art and music is imbued with awareness of our species' great objective of finding understanding of our corrupted condition—from Dorothy's *Somewhere Over The Rainbow* dream to 'wake up where the clouds are far behind me, where troubles melt like lemon drops' in *The Wizard of Oz*; to Schiller's words in Beethoven's *Ninth Symphony* of a time when the 'magic [of understanding] reunites…[and] all men become brothers…all good, all bad…be embraced, millions, this kiss [of reconciling understanding] for the

whole world'; to Christ's *Lord's Prayer* that '**Your** [integrative Godly] **kingdom come, your will** [that we be integrative] **be done on earth'**.

[160] And since everyone has intuitively been aware of the corrupted state of our human condition and task of finding understanding of it, it makes sense that everyone has also been intuitively aware that each individual, gender, race and culture is going to vary in how much they have been exposed to, and thus become adapted to, the upset state of the human condition—and therefore that individuals, genders, races and cultures are going to vary in how functional they are under the duress of the human condition. Some are going to be more innocent and thus naïve about life under the duress of the human condition, while others are going to be more adapted to, and thus more operational in, a human-condition-afflicted world, and still others are going to be overly adapted to that upset life and become cynical and opportunistic as a result. F. Essay 28 and chapters 8:16E-F of *FREEDOM* describe these differences in alienation and functionality between humans, especially between races, while chapter 8:11B describes the different roles of men and women in the human journey.

[161] And it also follows from this that everyone is going to be intuitively aware that those who are most functional are the 30 to 40-year-old equivalents who are not too naïve and not yet too cynical, which at this stage in the human journey are the so-called 'whites', especially white males. And being the most functional everyone is going to be intuitively aware that these white males are the ones who must especially undertake the task of finding understanding of the human condition.

[162] So what this means is that everyone has intuitively been waiting for these white male developers of the Judeo-Christian Western tradition to solve the human condition. And, further, since everyone has been intuitively aware of the critically important task this group has, everyone has been prepared to tolerate a degree of extra material good fortune and colonising power that the exceptional functionality afforded this group.

[163] BUT THIS TOLERANCE HAS ITS LIMIT: if this group is seen to be failing to deliver the answers that it was especially tasked with finding, then disenchantment with it, and resentment of its good fortune, would occur. And this is exactly what *has* occurred, the **'deepest hostility'** and **'denigration' 'of the Western tradition'** Murray wrote about.

[164] My favourite contemporary author, Sir Laurens van der Post, wrote about this situation as it existed in Africa, and beyond that, as it has existed in the whole world. In his 1955 novel *Flamingo Feather*, Sir Laurens wrote, **'I was old enough to remember some of the feeling of wonder and hope that the coming of the European had brought to many parts of Africa. The white man had been almost a kind of god to the African and, alas, subtly and fatally tempted as a result to exceed his common humanity. For generations the African had been happy to live in the hope of something better coming to him from the white man. But that hope was now running rapidly dry through our persistence in denying him his dignity and his own special capacity and honour as a human being...no human being could live indefinitely without honour and dignity...what was needed was someone, something big enough to straddle both** [heal the divide between blacks and whites]' (p.311-312 of 320).

[165] Through the characters in his novel, Sir Laurens then went on to write about how his white Dutch Boer ancestors had been **'always venturing courageously on and on into African's dark interior. [And wondered] Is this how the Great Trek for a better life is to end?...Is there no one great enough to take over the adventure and carry it on in some other dimension, to carry it on from world without to world within?** [the **world within** being the issue that had to be confronted and explained of our corrupted human condition]' (p.313). A few pages later he wrote more about this inner **'search which brings a man to the threshold of his private and personal task, the task that life demands of him day and night in his blood: to live with love out of love; to live the vision beyond reason or time which draws him from the centre of his being... To serve this vision, to protect it against all plausible substitutes, reasonable approximations and coward compromises is still, I believe, the knightly duty of contemporary man. If he shirks it I believe he shall never know inner peace.**

If a man accepts the challenge…of a cause beyond himself, then he has only to remain steadfast in pursuit of it and his life will achieve…something which is greater than happiness and unhappiness: and that is meaning' (pp.318-320).

[166] Yes, what Sir Laurens has recognised is that **'the knightly duty of'** the **'contemporary'** white male developers of the Judeo-Christian Western tradition has been **'to serve this vision, to protect it against all plausible substitutes, reasonable approximations and** [pseudo idealistic] **coward compromises…to remain steadfast in pursuit of…something which is greater than happiness and unhappiness: and that is meaning'**.

[167] So thank goodness **'the knightly duty of'** the white male upholders of the Judeo-Christian Western tradition has remained steadfast long enough to find that **'which is greater than happiness and unhappiness'**, namely the **'meaning'** of the human-condition-stricken **'world within'**. Condemned as being **'pale, stale males'**, the white male upholders of the Judeo-Christian Western tradition can finally be recognised as the heroes, legends and saviours it was always hoped they would be!

[168] And I should say that it wasn't just the relatively innocent, 20-year-old, or thereabouts, equivalent black Africans who became exasperated with the white male upholders of the Judeo-Christian Western tradition. The overly upset 50-year-old equivalent Arabs and Chinese, and the Russians living between them, have also become exasperated with the West's seeming failure to deliver the reconciling, redeeming and healing answers to the human condition, and as a result have hardened their pursuit of the old artificial substitute power, fame, fortune and glory ways of sustaining their sense of self-worth.

[169] I would like to make this point about China's very late appearance on the colonisation world stage with its global 'Belt and Road Initiative', which they assert is **'to enhance regional connectivity and embrace a brighter future'** (*Xinhua*, 28 Mar. 2015). Firstly, in paragraphs 1043-1044 of *FREEDOM* I explain that now that we can understand the human condition and acknowledge the various states of upset, we can admit that while "there is no doubt [the West's] colonialism had negative, exploitive repercussions, some truly terrible, like the slave trade",

"colonisation under the rule of 30-and-40-year-old equivalents did make significant sense. The 'races' who were most functional under the duress of the human condition tried to help the less functional become more materially developed—that is, advanced in the many arts of living with the human condition needed to progress humanity towards greater knowledge, ultimately understanding of the human condition. As Sir James Darling [my headmaster when I was at Geelong Grammar School, who is regarded as Australia's greatest ever educator] has written about the British Empire, **'the function of Empire is to educate rather than to oppress'**, and the [relatively isolated and thus relatively innocent] British have **'an unbeaten record in the history of civilization'** (*The Education of a Civilized Man*, ed. Michael Persse, 1962, pp.134, 136 of 223). Sir Laurens van der Post similarly wrote that **'Great Britain'** created **'the largest, the greatest and, I still believe, the best-organised, and most civilized empire in the history of the world'** (*The Admiral's Baby*, 1966, p.108 of 340)." What is significant about this is that being overly upset 50-year-old equivalents does mean that the Chinese form of colonisation is not likely to be primarily concerned with advancing "the many arts of living with the human condition needed to progress humanity towards greater knowledge, ultimately understanding of the human condition", like the Judeo-Christian Western tradition has been, but be more concerned with trying to salve an extremely insecure ego through winning power, fame, fortune and glory, a wrong emphasis that actually takes humanity *away* from rather than towards **'a brighter future'**. Colonialism practised properly was all about progressing the world towards the redeeming and reconciling understanding of the human condition, *not* about power and glory and domination. I might add that in 2022, China's President Xi Jinping reasserted his deluded and extremely dangerous vision when he claimed that his, and Russian President Putin's, authoritarian, thought-oppressing culture **'can calm a troubled world...play a guiding role to inject stability and positive energy into a world rocked by social turmoil'** (*The Australian*, 16 Sep. 2022).

[170] Yes, Singapore's former prime minister Lee Kuan Yew argued that **'Asian authoritarianism, rather than Western-style democracy, is necessary to produce the sort of economic growth East Asia is now experiencing'** (*The Sydney Morning Herald*, 14 May 1994), with the reason being, as he said in an interview, that **'democracy** [in Asia] **leads to undisciplined and disorderly conditions'** (*Swarajya* mag., 28 Mar. 2015). He even went on to claim that **'Democratic procedures have no intrinsic value. What matters is good government, whose primary duty is to create a stable and ordered society'** (ibid)! Certainly authoritarianism is effective in creating order amongst some overly upset 50-year-old equivalent Asian societies, and amongst relatively innocent 20-year-old or thereabouts black African societies (as has been done recently in Rwanda through Paul Kagame's leadership), but the truth remains that shutting down freedom of opinion, expression and individualism is actually an abandonment of humanity's fundamental task of searching for knowledge, ultimately for self-knowledge, the understanding of the human condition needed to bring about a truly ordered, happy, secure world—which Lee Kuan Yew actually partially acknowledged when he wrote about China that **'its creativity may never match America's because its culture does not permit a free exchange and contest of ideas'** (*TIME* mag., 4 Feb. 2013). The **'primary duty'** of the human race is actually NOT **'to create a stable and ordered society'**, but to find enough knowledge to end the human condition and human suffering forever, which is what the Judeo-Christian Western tradition has been tasked with doing and has now completed.

[171] It again needs to be emphasised that while we couldn't explain and defend the upset state of the human condition it was far too dangerous to acknowledge differences in alienation between people—because it only led to the more innocent being prejudiced towards the more upset who to them seem 'bad', and to the more upset retaliating with prejudice against the less upset for their either direct or implied condemnation of them. However, with the defence of upset found, it finally becomes both safe AND NECESSARY—if we

are to truly understand ourselves and free ourselves from living in Plato's horrible cave of soul-deadening darkness — to acknowledge differences in upset.

[172] Standing back and looking at our overall situation, the human race has been in an absolutely desperate situation. Antonio Gramsci certainly got it right when he said, **'The crisis consists precisely in the fact that the old is dying and the new cannot be born** [the human condition seemingly hasn't been able to be confronted and solved]; **in this interregnum a great variety of morbid symptoms appears.'** Clinical psychologist Maureen O'Hara also highlighted the horror of our plight when she said that **'humanity is either standing on the brink of "a quantum leap in human psychological capabilities or heading for a global nervous breakdown"'** (Richard Eckersley, address titled 'Values and Visions: Western Culture and Humanity's Future', Nov. 1995); as did the psychotherapist Wayne Dyer when he said, **'We've come to a place…where we can either destroy ourselves or discover our divineness'** (*The Australian Magazine*, 8 Oct. 1994); and as the journalist Doug Anderson also did when he wrote, **'Time may well be dwindling for us to enlighten ourselves…Tragic to die of thirst half a yard from the well'** (*The Sydney Morning Herald*, 31 Oct. 1994)! And finally, Benjamin Disraeli, a former Prime Minister of Great Britain, perfectly summarised the agony of our situation when he famously said, **'Stranded halfway between ape and angel is no place to stop.'** So it truly is an astronomical relief that the human race is no longer stranded in a hellish position and the redeeming explanation of our seemingly horribly flawed condition has finally been found!

CPSIA information can be obtained
at www.ICGtesting.com
Printed in the USA
BVHW062026310123
657534BV00021B/826